STUDENT SOLUTIONS MANUAL

TO ACCOMPANY

CALCULUS
FROM GRAPHICAL, NUMERICAL, AND SYMBOLIC POINTS OF VIEW

SECOND EDITION   **VOLUME 1**

OSTEBEE/ZORN

Arnold Ostebee
St. Olaf College

HOUGHTON MIFFLIN COMPANY   BOSTON   NEW YORK

Sponsoring Editor: Lauren Schultz
Editorial Associate: Marika Hoe
Senior Manufacturing Coordinator: Florence Cadran
Senior Marketing Manager: Michael Busnach

Copyright © 2002 by Houghton Mifflin Company. All rights reserved.

No part of this work may be reproduced or transmitted in any form or by any means, electronic or mechanical, including photocopying and recording, or by any information storage or retrieval system without the prior written permission of Houghton Mifflin Company unless such copying is expressly permitted by federal copyright law. Address inquiries to College Permissions, Houghton Mifflin Company, 222 Berkeley Street, Boston, MA 02116-3764.

Printed in the U.S.A.

ISBN: 0-618-25412-9

4 5 6 7 8 9 – MA – 07 06 05

TABLE OF CONTENTS

1 FUNCTIONS AND DERIVATIVES: THE GRAPHICAL VIEW
    1.1 Functions, Calculus Style   1
    1.2 Graphs   4
    1.3 A Field Guide to Elementary Functions   7
    1.4 Amount Functions and Rate Functions: The Idea of the Derivative   11
    1.5 Estimating Derivatives: A Closer Look   13
    1.6 The Geometry of Derivatives   16
    1.7 The Geometry of Higher-Order Derivatives   18
    Chapter Summary   21

2 FUNCTIONS AND DERIVATIVES: THE SYMBOLIC VIEW
    2.1 Defining the Derivative   24
    2.2 Derivatives of Power Functions and Polynomials   27
    2.3 Limits   30
    2.4 Using Derivative and Antiderivative Formulas   33
    2.5 Differential Equations; Modeling Motion   36
    2.6 Derivatives of Exponential and Logarithmic Functions; Modeling Growth   38
    2.7 Derivatives of Trigonometric Functions: Modeling Oscillation   41
    Chapter Summary   44

3 NEW DERIVATIVES FROM OLD
    3.1 Algebraic Combinations: The Product and Quotient Rules   46
    3.2 Composition and the Chain Rule   49
    3.3 Implicit Functions and Implicit Differentiation   51
    3.4 Inverse Functions and their Derivatives; Inverse Trigonometric Functions   53
    3.5 Miscellaneous Derivatives and Antiderivatives   56
    Chapter Summary   58

4 USING THE DERIVATIVE
    4.1 Slope Fields; More Differential Equation Models   61
    4.2 More on Limits: Limits Involving Infinity and l'Hôpital's Rule   63
    4.3 Optimization   66
    4.4 Parametric Equations, Parametric Curves   68
    4.5 Related Rates   70
    4.6 Newton's Method: Finding Roots   73
    4.7 Building Polynomials to Order; Taylor Polynomials   75
    4.8 Why Continuity Matters   77
    4.9 Why Differentiability Matters; The Mean Value Theorem   79

5 THE INTEGRAL
    5.1 Areas and Integrals   81
    5.2 The Area Function   85
    5.3 The Fundamental Theorem of Calculus   88
    5.4 Finding Antiderivatives; The Method of Substitution   90
    5.5 Integral Aids: Tables and Computers   94
    5.6 Approximating Sums: The Integral as a Limit   96
    5.7 Working with Sums   99
    Chapter Summary   104

APPENDICES

- A Machine Graphics — 109
- B Real Numbers and the Coordinate Plane — 111
- C Lines and Linear Functions — 114
- D Polynomials and Rational Functions — 116
- E Algebra of Exponentials and Logarithms — 118
- F Trigonometric Functions — 120
- G Real-World Calculus: From Words to Mathematics — 121

SECTION 1.1 FUNCTIONS, CALCULUS STYLE

§1.1 Functions, Calculus Style

1. (a) Reading from the altitude graph, $A(1) \approx 5700$ feet.

   (b) Reading from the velocity graph, $V(1) \approx 650$ feet/minute.

3. (a) The balloon was ascending at time $t = 1$ and descending at time $t = 6$.

   (b) Near time $t = 1$, the altitude graph is increasing; values of $A(t)$ get larger as one moves to the right from $t = 1$. Near time $t = 6$, the altitude graph is decreasing; values of $A(t)$ get smaller as one moves to the right from $t = 1$.

   (c) At time $t = 1$, the value of the $V(t)$ is positive. At time $t = 6$, the value of the $V(t)$ is negative.

5. The approximate population function is $p(t) = 906e^{0.008 \cdot (t-1800)}$. Recall that the function p is only an approximation—it can't give exact population figures. Here are the predictions in question:

   (a) For year 2010, we'd predict $p(2010) = 906e^{0.008 \cdot (2010-1800)} \approx 4861.19$ millions, or about 4.861 billion people.

   (b) For the year 1000, we get $p(1000) = 906e^{0.008 \cdot (1000-1800)} \approx 1.51$ million.

   (c) For the year 1000 B.C.E., use $t = -1000$. This gives the ridiculous prediction $p(-1000) = 906e^{0.008 \cdot (-1000-1800)} \approx 0.00000017$ millions, or about 0.17 people!

   (d) None of these answers are completely convincing. This isn't surprising, since we're using a simple formula to predict a complex phenomenon. Inputs near $t = 1800$ produce the most believable outputs.

7. In each part, the question is *which* line of the multiline definition of m to use.

   (a) $m(-4) = -3 \times (-4)/2 - 9/2 = 3/2 = 1.5$

   (b) $m(0) = -\sqrt{4 - (0+1)^2} = -\sqrt{3}$

   (c) $m(2.3) = 2.3 - 1 = 1.3$

   (d) $m(\pi) = -\pi + 5 \approx 1.8584$

9. (a) The circle with radius r and center at the point (a, b) can be described by the equation $(x-a)^2 + (y-b)^2 = r^2$. Thus, the circle with radius $r = 2$ and center at the point $(-1, 0)$ is described by the equation $(x+1)^2 + y^2 = 4$.

   (b) The equation for the circle found in part (a) can be written in the form $y^2 = 4 - (x+1)^2$. Thus, the equation $y = -\sqrt{4 - (x+1)^2}$ describes a portion of this circle. This is the equation that appears in the defintion of m. Thus, the graph of m over the interval $[-3, 1]$ is an arc of the circle of radius 2 centered at $(-1, 0)$.

11. Call the function f. Since the pieces of f are lines, the "formula" for f is

$$f(x) = \begin{cases} 3 & \text{if } -5 \leq x < -3 \\ -x - 2 & \text{if } -3 \leq x \leq 2 \\ 2x - 8 & \text{if } 2 < x \leq 5 \end{cases}$$

13. (a) The variable x is the edge length of the removed corners. Negative edge length doesn't make physical sense in the context of this example.

    (b) Since two edge lengths are removed from each side of the piece of paper, $2x$ must be less than the length of the shortest side of the paper. Thus, $2x < 8$ must be true for physical reasons.

15. (a) The volume is the product of the three dimensions: $V(w) = w(24 - 2w)(32 - 2w)$.

(b) The problem is to find, from among all legal inputs $0 \le w \le 12$, the one that produces the largest output $V(w)$. A graph of V for $0 \le w \le 12$ shows that the volume takes its largest value—a bit less than 1600 cubic inches—at $w \approx 4.5$. [NOTE: With calculus methods we could improve these guesses slightly, to get $w = (28 - \sqrt{208})/3 \approx 4.526$; $V \approx 1552.539$.]

17. (a) Since the area of a rectangle is width times height, $A(t) = 3t$.

    (b) The graph is a line segment joining the points (0, 0) and (5, 15).

19. (a) For each t, the area of the trapezoid is
$$A(t) = \frac{1 + (2t+1)}{2}(t) = \frac{t}{2}(2t+2) = t^2 + t.$$

    [NOTE: The area of a trapezoid is base $\times$ (ht$_1$ + ht$_2$)/2.]

21. (a) The domain of g is $[0, \infty)$. In other words, it is the set $\{w \mid w \ge 0\}$.

    (b) Yes — The range of g is $(-\infty, 12]$.

23. (a) 3 is in the domain of f.

    (b) 7 is in the range of f.

    (c) Yes — if $x = 3$, $2x + 1 = 7$.

    (d) Yes — if $t = 3$, $t^2 - 2 = 7$.

    (e) No — if $z = 3$, $z^3 - z^2 = 18 \ne 7$.

25. A function has a *unique* output for each input. Since, for example, the points $(x, y) = (0, -1)$ and $(x, y) = (0, 1)$ are on the graph — both with $x = 0$, the graph cannot be the graph of a function.

27. The domain of f is the set of inputs for which a value of the function is defined. Thus, the domain of f is the interval $[0, \infty)$. From the first two lines of the definition of f, it is apparent that the values 0 and -2 are in the range of f. From the third line of the definition, we can see that the interval $[2, \infty)$ is in the range of f. Thus, the range of f is $\{0, -2\} \cup [2, \infty)$

29. Because the domain of the square root function is $[0, \infty)$, the function f is defined if and only if $x^2 - 4 \ge 0$. Therefore, the domain of f is $(-\infty, -2] \cup [2, \infty)$.

31. The rule defining r makes sense as long as $(x+2)(1-x) > 0$. Now, the expression $(x+2)(1-x)$ can changes sign the points $x = -2$ and $x = 1$. Checking nearby values of x shows that $(x+2)(1-x) > 0$ only if $-2 < x < 1$. Therefore the natural domain of r is the interval $(-2, 1)$.

33. (a) $g(0) = 0$, $g(-1) = \sqrt{2} \approx 1.4142$, $g(1) = \sqrt{2} \approx 1.4142$, $g(2) = \sqrt{20} \approx 4.4721$, and $g(500) = \sqrt{62500250000} \approx 250000.5$.

    (b) The formula for the distance between the points (x_1, y_1) and (x_2, y_2) is $\sqrt{(x_2 - x_1)^2 + (y_2 - y_1)^2}$. It follows that the formula for g is $g(x) = \sqrt{x^2 + f(x)^2} = \sqrt{x^2 + x^4} = |x|\sqrt{1 + x^2}$.

35. (a) Specific values of j can be found by computing slopes directly: $j(2) = 3$, $j(1.1) = 2.1$, $j(1.01) = 2.01$, $j(0.99) = 1.99$, $j(0.9) = 1.9$, and $j(0) = 1$.

    (b) $x = 1$ is not in the domain of j because then the definition doesn't make sense. (A single point doesn't determine a line.)

    (c) The following formula defining $j(x)$ holds for all $x \ne 1$:
$$j(x) = \frac{f(x) - 1}{x - 1} = \frac{x^2 - 1}{x - 1} = \frac{(x+1)(x-1)}{x - 1} = x + 1.$$

    (d) The graph of j is the line $y = x + 1$ with a "hole" at $x = 1$.

SECTION 1.1 FUNCTIONS, CALCULUS STYLE

37. The rental company charges $30 for the first 100 miles, and $0.07 per mile for each mile over 100 miles.

39. (a) Volume = length × width × depth. Since the tank is a cube, $L = W = 10$ feet. Thus, the volume of the tank is $V = 100d$, where d is the depth of the water in the tank.

   (b) The *domain* is the interval of sensible depth values, so $0 \le d \le 10$ is the domain of V. (Depth can't be negative; depth over 10 feet overflows the tank.)

   (c) Plugging in $d = 0$ and $d = 10$ shows that the *range* of possible volumes (in cubic feet) is the interval $[0, 1000]$.

   (d) $V = 100d \implies d = V/100$. The latter expresses d as a function of V.

41. (a) The absolute value function is defined for all real numbers x. Thus, the domain of f is the interval $(-\infty, \infty)$. The range of f is the set of nonnegative real numbers; the interval $[0, \infty)$.

   (b) The slope of the f-graph is not defined at $x = 0$ because it has a sharp corner there. (Elsewhere, its slope is ± 1.)

   (c) Since the f-graph has a slope for all $x \ne 0$, the domain of g is $(-\infty, 0) \cup (0, \infty)$. Since the slope of the f-graph, when it is defined, is always ± 1, the range of g is the set $\{-1, 1\}$.

   (d) $g(x) = \begin{cases} -1 & \text{if } x < 0 \\ 1 & \text{if } x > 0 \end{cases}$

§1.2 Graphs

1. (a) The rule defining R makes sense if $t \neq 0$. Therefore, the domain of R is $(-\infty, 0) \cup (0, \infty)$.

   (b) The expression $1/t^2 > 0$ for all $t > 0$, so the range of R is $(0, \infty)$.

   (c) No, because 0 is not in the domain of R.

   (d) Yes; if $-5 < a < b < -1$, then $R(a) < R(b)$.

   (e) No; if $1 < a < b < 5$, then $R(b) < R(a)$ (i.e., R is decreasing on this interval).

3. B has a root at x if $B(x) = 0$. Thus, since the graph of B crosses the x-axis at $x = -3$, B has a root at $x = -3$.

5. The number x is a root of a function f if and only if $f(x) = 0$. Since $u = \sqrt{2}$ is a root of the function Y, $Y(\sqrt{2}) = 0$.

7. The function g is neither even nor odd. To see this, note that $g(1) = 2$ and $g(-1) = 0$. Thus, there is an x in the domain of g for which $g(-x) \neq \pm g(x)$; this implies that g is neither even nor odd.

9. No; it is apparent from a graph that g is concave down over a portion of the itnerval $[-2, 2]$.

11. (a) $f(x) = (x - 4)^3$ has the desired properties.

    (b) W has an inflection point at $x = 4$ because its graph changes its direction of concavity at this point.

13. The graph shows that I has a local maximum near $t = 12$ (i.e., on day 12). This means that the number of individuals with measles is largest at this time.

15. The inflection point in the graph of $S(t)$ near $t = 7$ means that the rate of change of the susceptible population is slowing down (because the concavity changes from concave down to concave up).

17. (a) The graph of j can be obtained from the graph of f by "stretching" the graph of f by a factor of 2.

    (b) The graph of k can be obtained from the graph of f by "stretching" the graph of f by a factor of $1/2$ (i.e., by compressing it).

    (c) The graph of m can be obtained from the graph of f by "stretching" the graph of f by a factor of 2, then reflecting it about the x-axis.

19. If x and $-x$ are both in the domain of f, then $f(x) = f(-x)$ since f is an even function. Thus, $f(-1) = f(1) = 2$.

21. Since T has period 7, $T(x + 7) = T(x)$. Thus, $T(7) = T(0 + 7) = T(0) = 3$.

23. Since T has period 7, $T(x + k \cdot 7) = T(x)$ for any integer k. Thus,
    $T(-30) = T(-30 + 10 \cdot 7) = T(-30 + 70) = T(40)$.

25. (a) $U(1.4) = T(7) = T(0) = 3$, since T has period 7.

    (b) Suppose that $U(x) = U(x + P)$. Then, $U(x + P) = T(5(x + P)) = T(5x + 5P) = T(5x) = U(x)$ if $5P = 7$. Since $U(x + 7/5) = U(x)$, U is periodic with period $7/5 = 1.4$.

27. Lines with positive slope go "up and to the right". Here, lines A, B, and C have positive slopes; lines D and E have negative slope. The slope of E is most negative; the slope of A is most positive. Thus, $m_E < m_D < 0 < m_B < m_C < m_A$.

29. From the graph, it appears that the lowest point of f over the interval $[-3, 3]$ is $(-1.5, -1.8)$. Thus, the minimum value of f over the interval is -1.8; it is achieved at $x = -1.5$.

31. No, because the graph shows that f is decreasing over part of this interval.

33. Yes. The sine function is one possible example.

SECTION 1.2 GRAPHS

41. The information given implies that $f(-3) = -5$. Since f is an even function, it follows that $f(3) = -5$. Thus, the point $(3, -5)$ is also on the graph of f.

43. (a) $g(x) = x^3 + 2$

    (b) The graph of g can be obtained by shifting a graph of f up by 2 units.

45. (a) $g(x) = 2x^3$

    (b) The graph of g can be obtained by stretching a graph of f vertically by a factor of 2.

47. The function is odd. Its graph is symmetric with respect to the origin.

49. The function is neither even nor odd. It is not symmetric about the origin and it is not symmetric about the y-axis.

51. The line A has slope 2 and passes through the point $(2, 3)$. Therefore, it can be described by the equation $y = 2(x - 2) + 3 = 2x - 1$.

53. Line A is described by the equation $y = 2x - 1$ and line B is described by the equation $y = 5 - x$. At $x = 4$, line A has height 7 and line B has height 1. Since the graph $y = g(x)$ lies between the lines A and B at $x = 4$, $1 < g(4) < 7$. Thus, $L = 1$ and $U = 7$.

55. (a) When D = water depth is 0, there is no water in the tank; hence $V(0) = 0$.

    (b) The tank is half of a sphere of radius 10 ft. Thus, when $D = 10$ ft, the tank is full of water. So, $V(10)$ is half the volume of a sphere of radius 10 ft. Thus, $V(10) = \frac{1}{2}\left(\frac{4}{3}\pi(10)^3\right)$ ft^3 = $\frac{2000\pi}{3}$ ft^3. [NOTE: The formula for the volume of a sphere of radius r is $\frac{4}{3}\pi r^3$. (Be careful, r is not the water depth!)]

    (c) When $D = 0$, $V = 0$. Also, the shape of the tank means that the volume increases faster and faster with respect to depth. Only Graph C has these properties.

57. If $C(t)$ is the cost of health insurance at time t, then any correct graph of C should be *increasing and concave up*.

59. The point $(4, 5)$ is another local maximum on the graph of F. Since F is periodic with period 7, $F(x) = F(x + 7)$. Therefore, $5 = F(-3) = F(7 - 3) = F(4)$. Since $(-3, 5)$ is a local maximum of F, $(4, 5)$ is also a local maximum.

61. (a) $g(x) = 2x^2 - 4x + 5 = 2\left(x^2 - 2x + 1\right) + 3 = 2(x - 1)^2 + 3 = 2f(x - 1) + 3$

    (b) The graph of g can be obtained from the graph of f by "stretching" vertically by a factor of 2, translating to the right by 1 unit, and then translating upward by 3 units.

63. (a) Yes, a periodic function can be even. The cosine function is an example.

    (b) Yes, a periodic function can be odd. The sine function is an example.

    (c) Yes, a periodic function can be neither even nor odd. The function $f(x) = 1 + \sin x$ provides an example.

65. Since the graph of f is symmetric with respect to the origin, f is an odd function. Therefore, $f(-1) = -f(1) = -3$.

67. (a) $y = f(a) + \dfrac{f(b) - f(a)}{b - a}(x - a)$ is the equation of the secant line through the points $(a, f(a))$ and $(b, f(b))$. Therefore, every point on the line segment between $x = a$ and $x = b$ lies above the graph of f.

    (b) $f(a) + \dfrac{f(b) - f(a)}{b - a}(x - a) > f(x) \implies \dfrac{f(b) - f(a)}{b - a} > \dfrac{f(x) - f(a)}{x - a}$.

    (c) The line segment lies below the graph of g.

69. (a) Let ℓ be the linear function that passes through the points $(1, 4)$ and $(5, 2)$ on the graph of g; an equation for this line is $\ell(x) = (9 - x)/2$. Since g is concave down on the interval $[0, 10]$, $g(3) > \ell(3) = 3$.

(b) Proceeding as in part (a) leads to the result $g(4) > 5/2$.

§1.3 A Field Guide to Elementary Functions

1. The expression involves a fractional power (i.e., $x^{1/2}$) rather than just positive integer powers.

3. The expression $(1 + 2x)/4x^3 = 0.25x^{-3} + 0.5x^{-2}$ involves negative powers of the variable.

5. Yes, because $(x^3 + 1)/(x + 1) = x^2 - x + 1$ for all $x \neq -1$.

7. (a) The natural domain of any polynomial is $(-\infty, \infty)$.

   (b) Since Q is a 6th-degree polynomial, it can have at most 6 roots.

   (c) Any function can have at most one y-intercept (i.e., if $x = 0$ is in the domain of the function, it must have a unique output value). Since Q is a function that includes 0 in its domain, it has one y-intercept.

   (d) Since Q is an even-degree polynomial, it is possible that $Q(x) > 0$ for all x. (For example, $Q(x) = x^6 + 1$ has no real roots.)

   (e) Since Q is an even-degree polynomial, it is possible that $Q(x) < 11$ for all x. (For example, $Q(x) = -x^6$ has this property.)

   (f) No. Every polynomial p has the property that $|p(x)| \to \infty$ as $x \to \pm\infty$.

9. $m(x) = \dfrac{3x}{x^2 + 1} + \dfrac{4}{5x + 6} = \dfrac{3x(5x + 6) + 4(x^2 + 1)}{(x^2 + 1)(5x + 6)} = \dfrac{19x^2 + 18x + 4}{5x^3 + 6x^2 + 5x + 6}$. Thus, $p(x) = 19x^2 + 18x + 4$ and $q(x) = 5x^3 + 6x^2 + 5x + 6$.

11. π^x involves a variable power of a fixed base, so it is an exponential function. x^π involves a fixed power of a variable base, so it is not an exponential function.

13. If $b > 0$, $b^0 = 1$. Therefore, the point $(0, 1)$ is on the graph of $y = b^x$ for every positive number b.

15. (a) All real numbers are in the natural domain of f. That is, the domain of f is the interval $(-\infty, \infty)$.

    (b) The range of f is all positive real numbers. That is, the range of f is the interval $(0, \infty)$.

    (c) f is increasing everywhere on its domain: if $a < b$, then $f(a) < f(b)$.

    (d) Since f is increasing everywhere on its domain, it does not have any local maximum or local minimum points.

    (e) f is concave up everywhere on its domain.

    (f) Since f is concave up everywhere, there are no points at which its concavity changes (i.e., f has no inflection points).

    (g) Since $f(x) > 0$ for all x, f has no roots.

17. (a) The domain of the cosine function includes all real numbers, so the domain of f is $(-\infty, \infty)$.

    (b) The range of the cosine function is the interval $[-1, 1]$, so the range of f is $[-1, 1]$.

    (c) Yes, f is an even function because the cosine function is an even function: $f(-x) = \cos(-x/2) = \cos(x/2) = f(x)$.

    (d) Yes, f is periodic with period 4π: $f(x + 4\pi) = \cos((x + 4\pi)/2) = \cos(x/2 + 2\pi) = \cos(x/2)$.

    (e) f is decreasing throughout the interval $(0, 2\pi)$. Therefore, f is increasing nowhere in the interval $(0, 2\pi)$.

    (f) Since f is decreasing throughout the interval $(0, 2\pi)$, f has no local maximum or local minimum points in this interval.

    (g) By examining a graph of f, it can be seen that f is concave up on the interval $(\pi, 2\pi)$. [NOTE: This is because the cosine function is concave up on the interval $(\pi/2, \pi)$.]

    (h) f has an inflection point at $x = \pi/2$ because the concavity of f changes from down to up at this point.

    (i) Since $f(\pi) = \cos(\pi/2) = 0$, f has a root at $x = \pi$. This is the only root of f in the interval $(0, 2\pi)$.

19. (a) The tangent function is not defined at x if $\cos x = 0$. Thus, the domain of f is $(-\infty, \infty)$ except $x = \pi/2 + k\pi$, where $k = 0, \pm 1, \pm 2, \ldots$.

   (b) The range of the tangent function includes all real numbers; it is the interval $(-\infty, \infty)$.

   (c) $\tan(-x) = \dfrac{\sin(-x)}{\cos(-x)} = \dfrac{-\sin x}{\cos x} = -\tan x$. Thus, $f(x) = \tan x$ is an odd function.

   (d) Yes, f is periodic and has period π because $\tan(x + \pi) = \dfrac{\sin(x + \pi)}{\cos(x + \pi)} = \dfrac{-\sin x}{-\cos x} = \dfrac{\sin x}{\cos x} = \tan x$.

   (e) f is increasing everywhere it is defined. Thus, f is increasing on the intervals $(0, \pi/2)$, $(\pi/2, 3\pi/2)$, and $(3\pi/2, 2\pi)$.

   (f) Since f is increasing everywhere it is defined, it has no local maximum or local minimum points.

   (g) f is concave up on the intervals $(0, \pi/2)$ and $(\pi, 3\pi/2)$.

   (h) f changes from concave down to concave up at $x = \pi/2$, $x = \pi$, and at $x = 3\pi/2$. Thus, these are inflection points of f.

   (i) Since $\tan(\pi) = 0$, f has a root at $x = \pi$.

   (j) f has vertical asymptotes at $x = \pi/2$ and at $x = 3\pi/2$.

21. (a) If $b > 0$, the exponential function b^x is defined for all real numbers x. Thus, the natural domain of h is the interval $(-\infty, \infty)$.

   (b) If $b > 0$, then $0 < b^x$. Thus, the range of h is the interval $(0, \infty)$.

   (c) Since $0 < b < 1$, the exponential function b^x is decreasing everywhere. Thus, there is no interval on which h is increasing.

   (d) Since h is decreasing throughout its domain, h has no local maximum or local minimum points.

   (e) For any $b > 0$, the exponential function b^x is concave up everywhere. Therefore, h is concave up on the interval $(-\infty, \infty)$.

   (f) Since h is always concave up, it has no inflection points.

   (g) Since $h(x) > 0$ for all x, h has no roots.

23. (a) For any $b > 0$, the logarithm function $\log_b x$ is defined for all $x > 0$. Thus, the natural domain of H is the interval $(0, \infty)$.

   (b) For any $b > 0$, the range of the logarithm function $\log_b x$ is the interval $(-\infty, \infty)$. Thus, the range of H is the interval $(-\infty, \infty)$.

   (c) If $0 < b < 1$, the logarithm function $\log_b x$ is decreasing throughout its domain. Therefore, H is not increasing anywhere.

   (d) Since H is decreasing throughout its domain, H does not have any local maximum or local minimum points.

   (e) If $0 < b < 1$, the logarithm function $\log_b x$ is concave up throughout its domain. Therefore, H is concave up on the interval $(0, \infty)$.

   (f) Since H is concave up throughout its domain, it has no inflection points.

   (g) For any $b > 0$, $\log_b 1 = 0$. Thus, H has exactly one root, at $x = 1$.

25. The domain of the cosine function is all real numbers; its range is the interval $[-1, 1]$. Therefore, the domain of f is the interval $(-\infty, \infty)$ and its range is $[-1, 1]$. [NOTE: The graph of f can be obtained from a graph of the cosine function by horizontal shifting and horizontal compressing.]

27. The domain of the exponential function e^u is $(-\infty, \infty)$. Thus, the domain of f is also $(-\infty, \infty)$. Since $0 \le x^2$ for all x, $1 \le e^{x^2}$ for all x. Therefore, the range of f is the interval $(-\infty, -1)$.

Section 1.3 A Field Guide to Elementary Functions

29. Both the sine and the natural exponential functions include all real numbers in their domain. Therefore, the domain of f is the interval $(-\infty, \infty)$. Since the range of the sine function is the interval $[-1, 1]$, $e^{-1} \leq e^{\sin x} \leq e^1$ for all x. This implies that the range of f is the interval $[e^{-1}, e]$.

31. The domain of the natural logarithm function is all positive real numbers, so x is in the domain of f if $x^3 + 1 > 0$. The latter condition will be true if $x > -1$, so the domain of f is the interval $(-1, \infty)$. Since the expression $x^3 + 1$ takes on all real values greater than 0 as x ranges through the interval $(-1, \infty)$, the range of f is the same as the range of the natural logarithm function. That is, the range of f is the interval $(-\infty, \infty)$.

33. Since the sine function is defined for all x, and since $2 + \sin x > 0$ for all x, the domain of f is the interval $(-\infty, \infty)$. Since the range of $2 + \sin x$ is the interval $[1, 3]$, the range of f is the interval $[3^{-1/3}, 1^{-1/3}] = [3^{-1/3}, 1^{1/3}]$.

35. The graph of f is the graph of the sine function shifted horizontally to the right by one unit. Since the sine function has period 2π, f also has period 2π.

37. The graph of f can be obtained by horizontally stretching the graph of the sine function by a factor of 2. Since the sine function has period 2π, f has period 4π.

39. The graph of f can be obtained by horizontally stretching the graph of the sine function by a factor of π. Since the cosine function has period 2π, f has period $2\pi^2$.

    Here's a symbolic proof of this result:
    $f(x + 2\pi^2) = \cos\big((x + 2\pi^2)/\pi\big) = \cos(x/\pi + 2\pi) = \cos(x/\pi) = f(x)$.

41. (a) For any $b > 0$, the natural domain of the logarithm function $\log_b x$ is the interval $(0, \infty)$. Thus, this interval is the domain of g.

    (b) Since $\log_{10} x = \dfrac{\ln x}{\ln 10}$, $g(x) = f(x)/\ln 10$ (i.e., $k = 1/\ln 10$).

    (c) The graph of g can be obtained by vertically stretching the graph of f by a factor of $1/\ln 10 \approx 0.43429$ (i.e., by compressing the graph of f vertically).

43. $e^B = e^{2 \ln A} = \left(e^{\ln A}\right)^2 = A^2$

45. If the point (A, B) is on the graph of the exponential function b^x, then $B = b^A$. This, in turn, implies that $\log_b B = A$. In other words, the point (B, A) is on the graph of the logarithm function $\log_b x$.

47. (a) The graph $y = g(x)$ is the graph $y = f(x)$ shifted vertically upwards by 3 units. Since the line $y = 2$ is a horizontal asymptote of f, this means that the line $y = 5$ is a horizontal asymptote of g.

    (b) The line $y = 2$ is a horizontal asymptote of g. Since the graph of g is a horizontal translation of the graph of f, both graphs have the same horizontal asymptotes.

    (c) The line $y = 7$ is a horizontal asymptote of g. Here's why: The function $f(x + 4)$ has the line $y = 2$ as a horizontal asymptote since f does. This implies that the function $3f(x + 4)$ has the line $y = 3 \cdot 2 = 6$ as a horizontal asymptote. Therefore, the function $3f(x + 4) + 1$ has the line $y = 6 + 1 = 7$ as a horizontal asymptote.

49. (a) If $x \neq \pm 3$, $g(x) = \dfrac{1}{x + 3} = \dfrac{1}{x + 3} \cdot \dfrac{x - 3}{x - 3} = \dfrac{x - 3}{x^2 - 9} = f(x)$.

    (b) Since f is a rational function, it is defined for all values of x where its denominator is nonzero. Thus, the domain of f is $(-\infty, -3) \cup (-3, 3) \cup (3, \infty)$. The range of f is the set of values $f(x)$ if x is in the domain interval. Thus, the range of f is $(-\infty, 0) \cup (0, 1/6) \cup (1/6, \infty)$. [NOTE: The values 0 and $1/6$ are not in the range of f because $x = -3$ and $x = 3$ are not in the domain of f.]

    (c) Since g is a rational function, its domain includes all values of x for which its denominator is nonzero (i.e., $x \neq -3$). Therefore, the domain of g is $(\infty, -3) \cup (-3, \infty)$. The range of g is the set of values that $g(x)$ assumes as x ranges through the domain of g. Thus, the range of g is $(-\infty, 0) \cup (0, \infty)$.

(d) They are the same except at $x = 3$. f is undefined at $x = 3$, but $g(3) = 1/6$.

(e) Yes. The line $y = 0$ is a horizontal asymptote of f.

(f) Yes. The line $x = -3$ is a vertical asymptote of f.

§1.4 Amount functions and rate functions: the idea of the derivative

1. $P(2) = 100$ means that the car is 100 miles east of the reference point (the State Capitol on-ramp in Bismarck, North Dakota) at 2:00 p.m. on October 1, 2000.

3. $V(2) = 70$ means that at 2:00 p.m. on October 1, 2000, the car is traveling east at 70 miles per hour.

5. V is the rate function associated with P. That is, $V(t)$ is the rate of change of P at time t.

7. If P is measured in kilometers and time is measured in hours, then the rate of change of P is expressed in kilometers per hour.

9. The line tangent to the graph of $P(t)$ at $t = 2$ has slope $P'(2) = V(2) = 70$.

11. $B'(t)$ represents the rate at which the bank balance is changing at time t (in dollars per year).

13. (a) $\Delta g = g(x + \Delta x) - g(x) = 3(x + \Delta x) - 1 - (3x - 1) = 3\Delta x$

    (b) Since g is a linear function, $g'(x)$ is the slope of the line defined by g. Thus, $g'(x) = 3$.

15. $f'(-2) \approx 10$

17. (a) Since the rate of change of v is a constant (-9.8), v must be a linear function with slope -9.8. Thus, $v(t) = v_0 - 9.8t$, where v_0 is a constant.

    (b) The constant v_0 is the velocity of the object at time $t = 0$ (i.e., the object's initial velocity).

19. Just before the instant of time at which the height is greatest, the velocity is positive. Just after the instant of time at which the height is greatest, the velocity is negative. Since velocity changes continuously, its value must be zero at the instant of time at which the height is greatest.

21. The height of the ball at time t is $h(t) = h_0 + v_0 t - 16t^2$, where h_0 is the height of the ball at time $t = 0$ and v_0 is the ball's velocity at time $t = 0$. Since the ball was thrown from a height of 5 feet with an initial vertical velocity of 30 ft/sec, $h(t) = 5 + 30t - 16t^2$.

    The velocity of the ball at time t is $v(t) = v_0 - 32t$, where v_0 is the ball's initial vertical velocity. Since the initial vertical veclocity of the ball is 30 ft/sec, its velocity at time t is $v(t) = 30 - 32t$ ft/sec.

23. A car with a top speed of 10 miles per hour can't travel more than 10 miles in one hour. Thus, after 1 hour, the car can't have traveled more than 10 miles (i.e., it can't be more than 10 miles from its starting position).

25. (a) Since $h'(x) \leq -2$ for all $x \geq 0$, $h(5) - h(0) \leq (-2)(5 - 0) = -10$. Since $h(0) = 0$, this implies that $h(5) \leq -10$.

    (b) Since $h'(x) \leq -2$ for all $x \geq 0$, $h(3) - h(0) \leq (-2)(3 - 0) = -6$. Since $h(0) = 0$, this implies that $h(3) \leq -6 < 5$.

    (c) Let $h'(x)$ be the eastward velocity of a car at time x. The inequality $h'(x) \leq -2$ for all $x \geq 0$ means that the car's eastward velocity is no greater than -2 (i.e., the car is always moving west at a speed of at least 2 miles per hour). Thus, after five hours, the car must be at least 10 miles west of its position at time $t = 0$. Similarly, after three hours, the car must be at least 6 miles west of its initial position and, therefore, it must be west of a position 5 miles east of its starting position.

27. Since $f'(x) \leq 4$, the speed limit principle implies that $f(x) - f(0) \leq 4x$ if $x \geq 0$, so $f(x) \leq 4x + 2$ if $x \geq 0$. Thus, $f(1) \leq 6$. The speed limit principle also implies that $f(x) \geq 4x + 2$ if $x \leq 0$, so $f(-3) \geq -10$.

29. (a) The speed limit principle implies that $g(x) - g(0) \leq 3x$ if $x \geq 0$, so $g(x) \leq 1 + 3x$ if $x \geq 0$. The speed limit principle also implies that $g(x) - g(0) \geq 3x$ if $x \leq 0$, so $g(x) \geq 1 + 3x$ if $x \leq 0$.

    (b) The curve $y = g(x)$ lies above the line $y = 1 + 3x$ to the left of $x = 0$ and below this line to the right of $x = 0$.

41. (a) 100 days after January 1 (i.e., on April 10), the length of a day is increasing at a rate of 0.05 hours per day.

   (b) In the northern hemisphere, days get shorter between about June 21 and December 21. Thus, $H'(t) < 0$ if $171 \le t \le 354$.

43. Yes. $f'(-1)$ is the slope of the line tangent to f at $x = -1$. Comparing the graph of f near $x = -1$ with the line through the points $(-2, -3)$ and $(0, -1)$ makes it clear that that $f'(-1) > 1$.

45. (a) f is a constant function, so $f'(x) = 0$ for any x. Thus, $f'(11) = 0$.

   (b) $f'(x) = 0$

47. Since $f'(1) = 2$, the tangent line has slope 2. Since $f(1) = 5$, the tangent line passes through the point $(1, 5)$. Thus, an equation of the tangent line is $y = 2(x - 1) + 5$.

49. (a) $g'(4)$ is the slope of the curve $y = g(x)$ at $x = 4$. Thus, $g'(4) = -2$.

   (b) The slope of the tangent line is $g'(4) = -2$. Therefore, since the tangent line passes through the point $(4, 3)$, the tangent line is described by the equation $y = -2(x - 4) + 3 = -2x + 11$.

51. Let $W(t)$ be the size of the work force as a function of time t. Then, $W'(t)$ is the rate at which the size of the workforce is changing. Since the workforce is growing more slowly now than it was five years ago, $W'(t_{now}) < W'(t_{now} - 5)$.

53. (b) Since the units associated with T are $°$C and the units associated with t are minutes, the units associated with $T'(t)$ are $°$C/minute.

   (c) Yes. $T(5)$ is the temperature of the coffee 5 minutes after it was poured into the cup. Since the temperature of the room is $25°$C, $T(5) \ge 25 > 0$.

   (d) No. $T'(5)$ is the rate of change of the temperature of the coffee in the cup 5 minues after it was poured. Since the initial temperature of the coffee is greater than the temperature of the room, the coffee is cooling. Thus, $T'(5) < 0$.

   (e) $T(300)$ is the temperature of the coffee 300 minutes (i.e., 5 hours) after it was poured into the cup. Since the temperature of the room is $25°$C, this will be (approximately) the temperature of the coffee in the cup.

   (f) $T'(300)$ is the rate of change of the temperature of the coffee in the cup 5 hours after it was poured into the cup. After this amount of time, the temperature of the coffee will be constant (i.e., it will remain at room temperature).

55. $f'(x) = 1$ implies that f is a linear function of the form $f(x) = x + C$, where C is a constant. Since $f(0) = 2$, $f(x) = x + 2$.

57. $f'(x) = -3$ implies that f is a linear function of the form $f(x) = -3x + C$, where C is a constant. Since $f(0) = -2$, $f(x) = -3x - 2$.

59. Since the line is tangent to the graph of f at $(5, 2)$, this point must be on the graph of f. Hence, $f(5) = 2$. The slope of the line tangent to the curve $y = f(x)$ at $x = 5$ is $f'(5)$. Since the line passing through the points $(5, 2)$ and $(0, 1)$ has slope $(2 - 1)/(5 - 0) = 1/5$, $f'(5) = 1/5$.

61. No. For example, if $f(x) = x$ and $g(x) = x + 2$, then $f'(x) = g'(x) = 1$ but $f(x) \ne g(x)$ for all x.

63. Since $f'(x) \le 0$ if $2 < x < 7$, the speed limit principle implies that
$f(6) - f(3) \le 0 \cdot (6 - 3) = 0 \implies f(6) \le f(3) \implies f(3) \ge f(6)$.

SECTION 1.5 ESTIMATING DERIVATIVES: A CLOSER LOOK

§1.5 Estimating Derivatives: A Closer Look

1. $f'(1) \approx 2$

3. Since $f'(1) = 2$, an equation of the line tangent to f at $x = 1$ is $y = 2(x - 1) + 1 = 2x - 1$.

5. $f'(-1) = -2$ means that that at $x = -1$, y is decreasing twice as fast as x is increasing.

7. The line ℓ is tangent to the graph of f of $x = 1$. Thus, near $x = 1$, $\ell(x)$ and $f(x)$ are visually indistinguishable.

9. (a)

x	0.7	0.8	0.9	1.0	1.1	1.2	1.3
$f(x)$	0.49	0.64	0.81	1.00	1.21	1.44	1.69

    (b) $f'(1) \approx \dfrac{f(1.1) - f(1.0)}{1.1 - 1.0} = \dfrac{1.21 - 1.00}{0.1} = 2.1$

11. $f'(-1) \approx \dfrac{f(-1.001) - f(-1)}{-0.001} = \dfrac{1.002001 - 1.00}{-0.001} = -2.001$. Alternatively,
    $f'(-1) \approx \dfrac{f(-0.999) - f(-1)}{0.001} = \dfrac{0.998001 - 1.00}{0.001} = -1.999$. Thus, it seems reasonable to guess that $f'(-1) \approx -2$.

13. $f'(1.4) \approx \dfrac{f(1.42) - f(1.38)}{1.42 - 1.38} \approx \dfrac{0.98865 - 0.98185}{0.04} = 0.17$

15. (a)

x	1.37	1.38	1.39	1.40	1.41	1.42	1.43
$f(x)$	0.97991	0.98185	0.98370	0.98545	0.98710	0.98865	0.99010

    (b) $f'(1.4) \approx \dfrac{f(1.41) - f(1.40)}{0.01} \approx \dfrac{0.98710 - 0.98545}{0.01} = 0.165$. Alternatively,
    $f'(1.4) \approx \dfrac{f(1.39) - f(1.40)}{-0.01} \approx \dfrac{0.98370 - 0.98545}{0.01} = 0.175$. Thus, it is reasonable to guess that $f'(1.4) \approx 0.17$.

17. (a) For all $x > 0$, $f(x) = x$ so $f'(2) = 1$.
    (b) For all $x < 0$, $f(x) = -x$ so $f'(-3) = -1$.
    (c) $f'(0)$ does not exist because f is not locally linear at $x = 0$. The graph of f has a sharp corner at $x = 0$ so the graph of f does not have a well-defined slope there.

19. The graph of f is symmetric about the y-axis (i.e., f is an even function). The line tangent to f at the point $(1, 1)$ is described by the equation $y = 2x - 1$. The reflection of this line across the y-axis is the line described by the equation $y = -2x - 1$. By symmetry, this line is tangent to the graph of f at the point $(-1, 1)$. Since the slope of this line is -2, $f'(-1) = -2$.

21. The graph of f is symmetric about the origin (i.e., f is an odd function). The line tangent of f at the point $(-1, -1)$ is described by the equation $y = 3x + 2$. The reflection of this line about the origin is the line described by the equation $y = 3x - 2$. By symmetry, this line is tangent to the graph of f at the point $(1, 1)$. Since the slope of this line is 3, $f'(1) = 3$.

23. (a) The formula $f'(x) = 2x$ predicts that $f'(3) = 6$.
    (b) $f'(3) \approx \dfrac{f(3.001) - f(3)}{0.001} = \dfrac{9.006001 - 9}{0.001} = 6.001$. Alternatively,
    $f'(3) \approx \dfrac{f(2.999) - f(3)}{-0.001} = \dfrac{8.994001 - 9}{-0.001} = 5.999$. Both results are consistent with the prediction in part (a) that $f'(3) = 6$.

25. (a) $f'(0.5) \approx \dfrac{f(0.501) - f(0.5)}{0.001} \approx 1.65$

(b) Yes — $f(0.5) = e^{0.5} \approx 1.65 \approx f'(0.5)$.

27. (a)

x	-2	-1	0	1
$f'(x)$	6.0	2.0	-2.0	-2.0

x	2	2.5	3	4
$f'(x)$	0.0	1.0	2.0	3.5

29. (a) $f'(1/4) \approx 1.0$, $f'(1) \approx 0.5$, $f'(9/4) \approx 0.33$, $f'(4) \approx 0.25$, $f'(25/4) \approx 0.20$, and $f'(9) \approx 0.17$.

(b) $f'(x) = 1/2\sqrt{x}$

31. (a) $f'(-1) \approx -2.72$, $f'(0) \approx 1.00$, $f'(1) \approx -0.368$, $f'(1.5) \approx -0.223$, and $f'(2) \approx -0.135$.

(b) $f'(x) = -e^{-x}$

33. Yes; for every increase of 0.1 in x, y increases by 1.4. Thus, the points lie on the line $y = 14x - 45$.

35. (a) $f'(0) = 0$, $f'(1/2) = 3/4$, $f'(1) = 3$, $f'(3/2) = 27/4$, $f'(2) = 12$.

(b) Since f is an odd function, f' will be an even function. (Reflecting a tangent line through the origin does not change its slope.)

(c) $f'(x) = 3x^2$

37. (a) $f'(1/4) = -16$, $f'(1/2) = -4$, $f'(1) = -1$, $f'(2) = -1/4$, $f'(3) = -1/9$.

(b) f is an odd function, so f' is an even function. (Reflecting a tangent line about the y-axis produces a line whose slope is -1 times the slope of the original line.)

(c) $f'(x) = -1/x^2$

39. (a) $f'(0) \approx 0$, $f'(\pi/6) \approx -0.5$, $f'(\pi/4) \approx -0.71$, $f'(\pi/3) \approx -0.87$, $f'(\pi/2) \approx -1$, and $f'(\pi) \approx 0$.

(b) $f'(x) = -\sin x$

41. (a) This line has slope $m_a \approx 1$.

(b) This line has slope $m_b \approx 0.62$.

(c) This line has slope $m_c \approx 0.43$.

(d) $\ln e = 1$, $\ln 5 \approx 1.61$, and $\ln 10 \approx 2.3$. $m_a \approx 1/\ln e$, $m_b \approx 1/\ln 5$, and $m_c \approx 1/\ln 10$.

43. (a) The slope of the line tangent to the graph of $y = 2^x$ at $x = 0$ is ≈ 0.693; at $x = 1$, the slope is ≈ 1.386; and, at $x = 2$, the slope is ≈ 2.773. Thus, $y'(0)/y(0) \approx 0.693$; $y'(1)/y(1) \approx 0.693$; and, $y'(2)/y(2) \approx 0.693$. Thus, the constant of proportionality is, approximately, 0.693.

(b) $\ln 2 \approx 0.693$

45. (a) The slope of the line tangent to the graph of $y = 10^x$ at $x = 0$ is ≈ 2.30; at $x = 1$, the slope is ≈ 23.03; and, at $x = 2$, the slope is ≈ 230.3. Thus, $y'(0)/y(0) \approx 2.30$; $y'(1)/y(1) \approx 2.30$; and, $y'(2)/y(2) \approx 2.30$. Thus, the constant of proportionality is, approximately, 2.30.

(b) $\ln 10 \approx 2.30$

47. (a) The slope of the line tangent to the graph of $y = \log_2 x$ at $x = 1/2$ is ≈ 2.89; at $x = 2$, the slope is ≈ 0.721; and, at $x = 10$, the slope is ≈ 0.144. Thus, $y'(1/2)/(1/2)^{-1} \approx 1.44$; $y'(2)/2^{-1} \approx 1.44$; and, $y'(10)/10^{-1} \approx 1.44$. Thus, the constant of proportionality is, approximately, 1.44.

(b) $1/\ln 2 \approx 1.44$

SECTION 1.5 ESTIMATING DERIVATIVES: A CLOSER LOOK

49. (a) The slope of the line tangent to the graph of $y = \log_{10} x$ at $x = 1/2$ is ≈ 0.869; at $x = 2$, the slope is ≈ 0.217; and, at $x = 10$, the slope is ≈ 0.0434. Thus, $y'(1/2)/(1/2)^{-1} \approx 0.434$; $y'(2)/2^{-1} \approx 0.434$; and, $y'(10)/10^{-1} \approx 0.434$. Thus, the constant of proportionality is, approximately, 0.434.

    (b) $1/\ln 10 \approx 0.434$

51. The rate of growth of h at x is $h'(x)$. Using numerical zooming, $h'(-3) \approx 0.141$, $h'(0) \approx 0$, $h'(1) \approx -0.841$, and, $h'(2) \approx -0.909$.

53. Since $h'(2) = -3$, the slope of the line tangent to h at $x = 2$ is -3. Thus, $\ell(x) = -3(x - 2) + h(2)$ is an equation of this tangent line. If $x \to x + \Delta x$, $\ell(x + \Delta x) - \ell(x) = -3\Delta x$. Since $\ell(x)$ is a good approximation to h near $x = 2$, we conclude that h decreases by approximately $-3\Delta x$.

55. Since $f'(x) = 2x$ and $g'(x) = -2$, $f'(x) = g'(x)$ at $x = -1$.

57. By graphing f and g on the same set of axes, it appears that f and g have the same slope at $x = 1$. Numerical zooming confirms that $f'(1) \approx -2 = g'(1)$.

59. If f and g are plotted using the viewing window $[3, 3.2] \times [-0.1, 0.1]$, they are visually indistinguishable. This suggests that $f'(\pi) = g'(\pi) = -1$.

61. (a) Since g is 2π-periodic, $g'(9\pi/4) = g'(\pi/4 + 2\pi) = g'(\pi/4) = -\sqrt{2}/2$.

    (b) Since $g'(\pi/4) = -\sqrt{2}/2$, the slope of the line tangent to g at $x = \pi/4$ is $-\sqrt{2}/2$. Now, the cosine function is symmetric about the y-axis (i.e., it is an even function), so the line tangent to the cosine function at $x = -\pi/4$ is the reflection across the y-axis of the line tangent to g at $x = \pi/4$. This (reflected) line has slope $\sqrt{2}/2$. Thus, $g'(-\pi/4) = \sqrt{2}/2$.

    (c) By symmetry, $g(\pi/2 + x) = -g(\pi/2 - x)$ so $g'(\pi/2 + x) = -g'(\pi/2 - x)$. Substituting $x = \pi/4$ leads to $g'(3\pi/4) = -g'(\pi/2 - \pi/4) = -g'(\pi/4) = -\sqrt{2}/2$.

§1.6 The Geometry of Derivatives

1. (a) The tangent line is horizontal, so its slope is 0.

   (b) Yes. The point A' is on the x-axis, so it corresponds to an f' value of 0.

3. The function f achieves larger values elsewhere. For example, $f(3.2) \approx 5$ which is larger than the value of f at A.

5. If g has a local minimum point at 2, $g(2) \leq g(x)$ for all values of x in an interval containing $x = 2$.

7. If x_0 is a global maximum point of a function f, $f(x_0) \geq f(x)$ for every x in the domain of f. If x_0 is a local maximum point of a function f, the statement $f(x_0) \geq f(x)$ may not be true for every x in the domain of f (it only needs to be true for every x in an open interval containing x_0).

9. At a stationary point, the graph of a function has a flat spot (i.e., it has a horizontal tangent line).

11. Yes. f is decreasing to the left of $x = 3$ and increasing to the right of $x = 3$. Therefore, f has a local minimum point at $x = 3$.

13. An inflection point is a point at which a graph's direction of concavity changes.

17. f has a stationary point (i.e., a horizontal tangent line) between A and B, and also between C and D.

19. f' is inreasing between the points C and D because the graph of f is concave up between these two points.

21. (a) f' is negative where f is decreasing. Therefore, $f'(x)$ is negative on the intervals $(-5, -2)$ and $(2, 5)$.

    (b) f' is decreasing where f is concave down. Therefore, f' is decreasing on the interval $(0, 5)$

23. f' achieves its maximum value at the point where f changes its concavity from up to down (i.e., at an inflection point). This occurs at $x = 0$; $f'(0) \approx 1.4$.

25. g is increasing on those intervals where $g' \geq 0$. Therefore, g is increasing on the intervals $(0, 5.5)$ and $(9, 10)$.

27. g has stationary points where $g'(x) = 0$. Therefore, g has stationary points at $x = 1$, $x = 5.5$, and $x = 9$.

29. g has a local minimum where g' changes sign from negative to positive. Thus, $x = 9$ is a local minimum point for g.

31. No. Since g' is decreasing at $x = 6$, g is concave down at $x = 6$.

35. Since $g'(9) = 0$, the line tangent to g at $x = 9$ is a constant linear function. Since $g(9) = -1$, the tangent line is $y = -1$.

37. f is increasing on those intervals where $f' \geq 0$. Thus, f is increasing on the interval $(-3, 4)$.

39. f has a stationary point at x if $f'(x) = 0$. Thus, f has stationary points at $x = -3$ and at $x = 3$.

41. f has a local minimum where the sign of f' changes from negative to positive. Thus, f has a local minimum point at $x = -3$.

43. f has a point of inflection where f' has a local maximum or minimum. Thus, f has points of inflection at $x = -1$ and $x = 3$.

45. Since $f'(1) = 2$, the slope of the tangent line is 2. Since $f(1) = 3$, the tangent line passes through the point $(1, 3)$. Therefore, $y = 2(x - 1) + 3 = 2x + 1$ is an equation for the line tangent to f at $x = 1$.

47. (a) Roots of f' correspond to stationary points of f (i.e., places where the tangent line is horizontal). Since f has only one stationary point, at $x = 0$, this point is the only root of f'.

SECTION 1.6 THE GEOMETRY OF DERIVATIVES

    (b) Since f is increasing over the interval $(-\infty, 0)$, $f' > 0$ on this interval.

    (c) Yes. The graph of f is concave up at $x = 2$, so f' is increasing at $x = 2$.

    (d) $f'(x) = -2x/\left(1 + x^2\right)^2$

49. (a) Roots of f' correspond to stationary points of f (i.e., places where the tangent line is horizontal). Since f has no stationary points, f' has no roots.

    (b) The graph of f is increasing on the interval $(0, \infty)$, so f' is positive on this interval.

    (c) The graph of f is concave down on the interval $(0, \infty)$, so $f'' < 0$ on this interval. Therefore, f' is decreasing over the interval $(0, \infty)$.

    (d) $f'(x) = 1/x$

51. (a) The graph of f is increasing on those intervals where $f' \geq 0$. Therefore, f is increasing on the intervals $(-\infty, -4)$ and $(2, \infty)$.

    (b) x is a stationary point of f if $f(x) = 0$. Therefore, $x = -4$ and $x = 2$ are stationary points of f.

    (c) $x = a$ is a local minimum point of f if the sign of f' changes from negative to positive at $x = a$. Since $f'(x)$ changes sign from negative to positive at $x = 2$, f has a local minimum point there.

    (d) The graph of f is concave down on intervals where f' is decreasing. Thus, f is concave down on the interval $(-\infty, -1)$.

    (e) $f(x) = x(x^2 + 3x - 24)/3$ is one possibility.

53. (a) $f'(x) > 0$ for all x, so f is increasing on the interval $(-\infty, \infty)$.

    (b) The equation $f'(x) = 0$ has no solutions, so f has no stationary points.

    (c) Since $f'(x) > 0$ for all x, there is no point at which the sign of f' changes from negative to positive. Therefore, f has no local minimum points.

    (d) f' is decreasing on the interval $(0, \infty)$ so f is concave down on this interval.

55. $y = 0$

57. $y = -1/x$

59. $y = -2$

61. $g(x) = e^{-x}$

63. $k(x) = -e^x$

§1.7 The Geometry of Higher-Order Derivatives

1. (a) The graph of f is concave up at $x = -1$.

   (b) The graph of f' is increasing at $x = -1$.

3. (a) $h''(2) = 0$ and h'' changes sign at $x = 2$

   (b) The graph of h' has a local extreme point at $x = 2$.

5. Yes — the second derivative test implies that f has a local minimum at $x = -3$.

7. Stationary points of a function f are, by definition, roots of f'. Since f' is a quadratic function, it has only two roots so f has only two stationary points.

9. $f''(1) < f''(3) < f''(-2) < f''(-3)$ because f'' is the derivative of f'.

11. (b) Since g is concave down over the interval $[1, 5]$, the graph of f lies below the line tangent to g at $x = 2$ throughout this interval. In other words, $g(x) < \ell(x)$ if $1 < x < 2$ or $2 < x < 5$. Thus, in particular, $g(3) < \ell(3)$.

    (c) No. The argument in part (a) shows that $g(1.5) < \ell(1.5)$.

13. (a) f'' is positive where the graph of f is concave upward. Therefore, $f''(x)$ is positive if $-10 < x < 1$.

    (b) f'' is negative where the graph of f is concave downward. Therefore, $f''(x)$ is negative if $1 < x < 10$.

    (c) f'' is zero where the graph of f changes concavity. Therefore, $f''(x) = 0$ only for $x = 1$.

15. g is concave up on intervals where $g'' \geq 0$. Therefore, g is concave up on the interval $(0, 3)$.

17. No. g does not have a point of inflection at $x = 1$ because $g''(x)$ does not change sign at $x = 1$.

19. Since $g''(x) \geq 0$ throughout the interval $[0, 3]$, g' is increasing on this interval. This implies that $g'(0) < g'(1) < g'(2) < g'(3)$.

21. Yes. $g''(x) < 1$ if $1 \leq x \leq 3$. Therefore, the speed limit principle implies that $g'(3) - g'(1) \leq 1 \cdot (3 - 1) = 2$ so $g'(3) < -2$. Since $g'(3)$ is negative, g is decreasing at $x = 3$.

23. (a) Values of g'' can be estimated by measuring slopes of lines tangent to g'. The results are $g''(0.5) = 0.125$, $g''(1) = -1/2$, $g''(1.5) = 1.125$, and $g''(2) = 2$.

25. (a) This function is concave upward if $x < 0$, concave downward if $x > 0$, and its second derivative doesn't exist at $x = 0$. Thus, graph (iv) must be the second derivative function.

    (b) This function is concave downward everywhere, so its second derivative must be negative everywhere. Thus, graph (iii) is the second derivative function.

    (c) This function is concave down except on an interval centered at the origin where it is concave up. Thus, the second derivative is negative except on an interval centered at the origin where it is positive. It follows that graph (i) is the second derivative function.

    (d) Since the function is a straight line, its first derivative is a horizontal line so its second derivative is zero everywhere. Thus, graph (ii) is the second derivative function.

27. **B** is the graph of f, **A** is the graph of f', and **C** is the graph of f''.

    Looking at the behavior of the three graphs near $x = 0$, one sees that **A** must be the derivative of **B** (**B** is increasing until its local maximum point; **A** is positive on the interval where **B** is increasing; etc.) and that **C** must be the derivative of **A** (**C** has a root at each local extreme point of **A**).

29. (a) $V''(t)$ is the rate at which V' is changing at time t. Since water is flowing into the tank at a *constant* rate, $V''(t)$ is **zero**.

SECTION 1.7 THE GEOMETRY OF HIGHER-ORDER DERIVATIVES

(b) $H''(t)$ is the rate at which H' is changing at time t. As the tank fills, it takes more water (i.e., more time) to raise the height of the water in the tank by a fixed amount. Thus, H' is "large" when the tank is nearly empty and H' is "small" when the tank is nearly full. These considerations imply that $H''(t^*) < 0$.

31. Let $T(t)$ be the child's temperature at time t. $T'(t) > 0$ (that is, $T(t)$ is *increasing*), but $T''(t) < 0$ (that is, $T(t)$ is increasing at a *decreasing* rate.)

33. The function $f(x) = e^{-x}$ has the desired properties.

35. If f is an even function, then $g = f'$ is an odd function. This implies that $g' = f''$ is an even function.

37. (a) Since f is concave up everywhere, f' is increasing everywhere. This implies that f can have at most one root and the root, if it exists, must be located in the interval $(-\infty, 2)$.

    In fact, one can say more about the possible location of a root of f. Since f is concave up on the interval $(-\infty, 2)$, the curve $y = f(x)$ must always lie *above* the line tangent to f at $x = 2$. Since this line crosses the x-axis at $x = -1/2$, the root, if it exists, must be located in the interval $(-\infty, -1/2)$.

    (b) Since f is concave up on the interval $[0, 2]$, the line tangent to the graph of f at $x = 2$ lies *below* the curve $y = f(x)$. Thus, the y-intercept of the tangent line is a lower bound on the value of $f(0)$. Since the tangent line is $y = 2(x - 2) + 5 = 2x + 1$, we have that $f(0) \geq 1$.

    (c) Since f is concave up on the interval $[-2, 2]$, the line passing through the points $(-2, f(-2)) = (-2, -1)$ and $(2, f(2)) = (2, 5)$ must lie above the curve $y = f(x)$ over the interval $(-2, 2)$. The equation of this line is $y = \frac{3}{2}(x + 2) - 1 = \frac{3}{2}x + 2$. At $x = 0$ this line passes through the point $(0, 2)$, so $f(0) < 2$ must be true. This implies that $f(0) < 3$.

39. (a) $f''(-1) = 0$, $f''(0) = 2$, and $f''(2) = 4$.

    (b) f'' has a root at each stationary point of f'. Since f' has only one stationary point (a local minimum point at $x = -1$), f'' has only one root.

    (c) $f''(x) > 0$ on those intervals where f' is increasing. Thus, $f''(x) > 0$ on the interval $(-1, \infty)$.

    (d) $y = 2x + 2$

41. (a) $f''(-1) = -1/2$, $f''(0)$ does not exist, and $f''(2) = 1/3$.

    (b) f'' has a root at each stationary point of f'. Since f' has no stationary points, f'' has no roots.

    (c) $f''(x) > 0$ on those intervals where f' is increasing. Thus, $f''(x) > 0$ on the interval $(0, \infty)$.

    (d) $y = x/(|x| + x^2)$

43. (a) f is concave up on the intervals where $f''(x) > 0$. Thus, f is concave up on the interval $(2, \infty)$.

    (b) f' is decreasing on the the intervals where $f''(x) < 0$. Thus, f' is decreasing on the interval $(-\infty, 2)$.

    (c) f' has a stationary point at each root of f''. Thus, f' has two stationary points: at $x = -1$ and at $x = 2$.

    (d) f has an inflection point where the sign of f'' changes. Thus, f has an inflection point at $x = 2$.

    (e) $y = (x^4 - 6x^2 - 8x)/4$

45. (a) f is concave up on the intervals where $f''(x) > 0$. Thus, f is concave up on the interval $(-1, \infty)$.

    (b) f' is decreasing on the the intervals where $f''(x) < 0$. Thus, f' is decreasing on the interval $(-\infty, -1)$.

    (c) f' has a stationary point at each root of f''. Thus, f' has only one stationary point, at $x = -1$.

    (d) f has an inflection point where the sign of f'' changes. Thus, f has an inflection point at $x = -1$.

    (e) $y = \sqrt{1 + x^2} + \ln\left|x + \sqrt{1 + x^2}\right|$

47. (a) Yes. Since $f'(1) = 0$ and $f''(1) = 5 > 0$, f has a local minimum point at $x = 1$.

    (b) No. $f''(x) > 0$ if $-1 < x < 1$, so f is concave up over this interval. This implies that the curve $y = f(x)$ lies above the line tangent to f at $x = -1$ over the interval $-1 < x < 1$.

49. (a) No. Since $f'(1) = 0$ and $f''(1) = -1/6 < 0$, f has a local maximum point at $x = 1$.

    (b) Yes. $f''(x) < 0$ if $-1 < x < 1$, so f is concave down over this interval. This implies that the curve $y = f(x)$ lies below the line tangent to f at $x = -1$ over the interval $-1 < x < 1$.

51. $f''(x) = 3 + 2x$

§1.8 Chapter summary

1. In general, if f' is a constant function, then f is a linear function. Since $f'(x) = 3$, $f(x) = 3x + C$, where C is a constant.

3. (a) The graph of g is a vertical translate of the graph of f.
   (b) The graphs of g' and f' are identical (i.e., $g'(x) = f'(x)$ for every x in the domain of f'.
   (c) $g'(1) = f'(1) = 5$

5. Since h is a vertical translate of f, $h'(2) = f'(2) = 3$. Thus, the line $y = (3x - 4) + 1 = 3x - 3$ is an equation of the line tangent to h at the point $(2, h(2)) = (2, 3)$.

7. Since $g(x) = f(x - 2)$, the graph of g is a horizontal translate of the graph of f. This implies that $g'(x) = f'(x - 2)$. Thus, $g'(0) = f'(-2) = 1/7$.

9. The shape of the f-graph repeats every 5 units. Since f' is the slope function, it must also be periodic with period 5.

11. $f'(x)$ is the slope (or rate of growth) of f at x. From a graph of f it is apparent that $f'(-1) < f'(0) < f'(2) < f'(10)$.

13. For any function f, $f'(x)$ is the slope of f at x; it is also the rate of growth of f at that point. A close look at graphs of f and g near each of the listed points leads to the following conclusions: $f'(1/10) < g'(1/10)$, $f'(1/2) < g'(1/2)$, $g'(2) < f'(2)$, and $g'(7) < f'(7)$. Thus, g is growing faster at the first two points and f is growing faster at the other two points.

15. Neither graph **A** nor **B** can be f since then $f'(x) < 0$ would be true on the interval $[0, 1]$ and graph **C** does not have this property. Thus, graph **C** must be the graph of f. Estimating $f'(x)$ at some point (e.g., at $x = 2$, $f'(2) \approx -0.4$) allows us to determine that **A** is the graph of f'.

17. At $x = 3$ and at $x = 6$.

19. f is concave down wherever the slope of f' is negative. This occurs on the intervals $(1, 2)$ and $(4.5, 7)$.

21. Since $f'(x) < 0$ on the interval $[0, 2]$, f is decreasing there. Thus, f achieves its maximum value at $x = 0$ and its minimum value at $x = 2$.

23. Notice that f' is negative on the interval $(1, 4]$ and positive on the interval $[-2, 1)$. Therefore:

    (a) f is increasing on $(-2, 1)$
    (b) Since $f'(1) = 0$, $x = 1$ is a stationary point of f. It is a local maximum point since f' changes from positive to negative there (i.e., f changes from increasing to decreasing at $x = 1$).
    (c) The graph has more or less the shape of an inverted "vee," with vertex at $x = 1$ (e.g., something similar to the graph of $y = e^{-x^2/40}$).

25. (a) Since $f'(1) = 0$ and $f'(x)$ changes sign at $x = 1$, f has a local extreme point there.
    (b) No, $x = 1$ is a local minimum point of f since $f'(1) = 0$ and the sign of $f'(x)$ changes from negative to positive at $x = 1$.
    (c) No, $f'(-2) = 0$ but $f'(x)$ doesn't change sign at $x = -2$ so $x = -2$ is not a local extreme point of f.
    (d) f is increasing where $f'(x) \geq 0$. Thus, f is increasing on the interval $(1, \infty)$.
    (e) f is concave down where f' is decreasing. Thus, f is concave down on the interval $(-2, 0)$.
    (f) f has inflection points at each point where f' has a local extreme point. Thus, f has inflection points at $x = -2$ and $x = 0$.

27. (a) Since $f(2) = 4$ and $f'(2) = \sqrt{8+1} = 3$, the line tangent to f at $x = 2$ passes through the point $(2, 4)$ with slope 3. An equation of this tangent line $y = 3(x - 2) + 4 = 3x - 2$.

 (b) Since f' is an increasing function on the interval $[-1, \infty)$, f is concave up over this interval. This implies that the curve $y = f(x)$ lies above any tangent line in this interval. Thus, $\ell(0)$ underestimates $f(0)$.

 (c) Using the tangent line found in part (a), $f(2.1) \approx 4.3$. From part (b), we may conclude that $f(2.1) \geq 4.3$ (i.e., the tangent line estimate is too small).

29. Using the speed limit principle, $f(5) - f(1) \leq 3(5 - 1) \implies f(5) \leq f(1) + 12 = 10$.

31. Yes. Using the speed limit principle,
 $f(-5) - f(1) \geq 3\big((-5) - 1\big) \implies f(-5) \geq f(1) - 18 = -20 > -23$.

33. Using the speed limit principle, the condition $f'(x) \leq 3$ if $-10 \leq x \leq 10$ implies that $f(8) - f(1) \leq 3(8 - 1) \implies f(8) \leq f(1) + 21 = 19$. Thus it is possible that $f(8) = -25$.

35. **A** is the graph of f'', **B** is the graph of g, **C** is the graph of f', and **D** is the graph of f.

37. The function $f(x) = -e^{-x}$ has the desired properties.

39. No such function exists. The conditions given imply that f lies below the x-axis and that it is concave up everywhere (i.e., its slope is always increasing). Such a function must cross the x-axis if its domain is all real numbers.

41. Yes, it is easy to sketch a function with the desired properties.

43. No; since $x = 2$ is a local maximum point of g, $g''(2) \leq 0$ must be true.

45. Yes, the slope of g at $x = 3$ could be 1.

47. No — $f'(x) \geq 0$ throughout the interval $[-2, 3]$, so $f(3) > f(-2) = -3$.

49. Since $f'(x) \geq 2$ if $-2 \leq 0 \leq 0$, the speed limit principle implies that
 $f(0) - f(-2) \geq 2 \cdot \big(0 - (-2)\big) = 4 \implies f(0) \geq 1$ since $f(-2) = -3$.

51. f achieves its smallest value at $x = -2$. If $-2 \leq x \leq 2$, then $f'(x) > 0$ and the speed limit principle implies that $f(x) - f(-2) \geq 0 \implies f(x) \geq f(-2)$.

53. (a) Yes. The particle is moving up at $t = 1$ because $y(t)$ is increasing at $t = 1$.

 (b) Yes. $y'(1)$ is the slope of the curve $y(t) = \sin t$ at $t = 1$. Since the y-graph is increasing at $t = 1$, $y'(1) > 0$.

55. $f'(-2)$ is the slope of the tangent line, so $f'(-2) = 4$. Since the tangent line and the function f "touch" at $x = -2$, $f(-2) = 4 \cdot -2 + 3 = -5$.

57. No. The graph shows that $g'(x) > g'(2)$ if $2 < x \leq 5$. The tangent line has a constant growth rate, $g'(2)$. However, throughout the interval $(2, 5]$, the growth rate of g is always greater than $g'(2)$. Thus, the tangent line at $x = 2$ lies below the graph of g over the interval $(2, 5]$.

59. Yes. Since $g'(x) > 0$ if $3 \leq x \leq 5$, g is increasing over this interval. Thus, $g(3) < g(5)$.

61. In 1810, the population of the United States was growing at a rate of approximately 2.3 million people per year.

63. No — $g'(3) \approx -0.2781 < 0$ implies that g is *de*creasing at $r = 3$.

65. No — g' is a decreasing function at $r = 2$ so g is concave down there.

SECTION 1.8 CHAPTER SUMMARY

67. g has one local minimum point in the interval $[0, 5]$, at $r = \sqrt{10\pi/3} \approx 3.326$. This is the only point in the interval where the sign of g' changes from negative to positive.

69. Local extrema of g' correspond to inflection points of g. Since g' has a local minimum at $r = \sqrt{2\pi} \approx 2.5066$, g has an inflection point there.

71. The graph of g' is decreasing throughout the interval $[1, 1.1]$, so the graph of g is concave down on this interval. It follows that the line tangent to the graph of g at $r = 1$ lies above the graph of g over the interval $[1, 1.1]$.

73. (a) $h'(z) = 3 \implies h'(1) = 3$. (Since h is a linear function, h' is a constant.)

    (b) Since h' is a constant, $h''(z) = 0$ for any z. Therefore, $h''(4) = 0$.

75. A "qualitatively correct" graph of G must be (i) increasing on the interval $[1, 3]$, (ii) concave up on the interval $[1, 2]$, (iii) concave down on the interval $[2, 3]$, and (iv) have an x-intercept at $x = 1$. The curve $y = 6x^2 - x^3 - 5$ has the right shape.

77. $f'(x)$ is the slope of f at x. Using this idea, it is clear from the graph that $-2 < f'(1.5) < f'(3) < 0 < 0.5 < f'(4.5)$.

79. No. The maximum value of f must occur at a stationary point (i.e., at a root of f') or at an endpoint of I. Thus, the maximum value of f on I *cannot* occur at 4.

83. (a) No — $h'(1) = -2 < 0$ so h is decreasing at $w = 1$.

    (b) Yes — h is concave up at $w = 4$ because $h'(w)$ is increasing at $w = 4$.

    (c) The point $w = 9$ is a local minimum point because the sign of h' changes from negative to positive there.

85. $g(4) = -7$. Since g' is a constant function, g must be a linear function. From the information given, it follows that $g(t) = -3(t - 1) + 2 = 5 - 3t$.

87. The statement *must* be true because $G''(w)$ changes sign at $w = -2$.

89. The statement *might* be true. G is concave up on the interval $[0, 2]$ (since $G''(w) > 0$ if $0 \leq w \leq 2]$), but G need not be decreasing over this interval.

91. (a) At the time of the election, 2 million people were unemployed.

    (b) Two months after the election, 3 million people were unemployed.

    (c) Twenty months after the election, unemployment was increasing at a rate of 10,000 people per month.

    (d) Three years (thirty-six months) after the election, the number of unemployed was at a local minimum.

93. (a) g has an inflection point at $w = \sqrt{\pi}$ because this point corresponds to a local maximum of $g'(w)$.

    (b) Yes, g also has an inflection point at $w = \sqrt{2\pi}$ because this point corresponds to a local minimum of $g'(w)$.

95. g is concave down on the interval $(\sqrt{\pi}, \sqrt{2\pi})$ because the graph of g' is decreasing on this interval.

97. g has a local maximum point where the sign of g' changes from positive to negative. This occurs once in the interval $[1, 3]$, at $w = \sqrt{15\pi/3}$.

99. (d) When the car has positive acceleration, the graph in part (a) is concave up, the graph in part (b) is increasing, and the graph in part (c) is positive.

    When the car has zero acceleration, the graph in part (a) is linear, the graph in part (b) is horizontal, and the graph in part (c) is zero.

    When the car has negative acceleration (i.e., it is decelerating), the graph in part (a) is concave down, the graph in part (b) is decreasing, and the graph in part (c) is negative.

§2.1 Defining the Derivative

1. $D'(2) \approx \dfrac{D(2.05) - D(1.95)}{2.05 - 1.95} = \dfrac{83.54 - 76.54}{0.1} = 70$ mph

3. (a) The slope of M is $\dfrac{f(2.005) - f(2)}{0.005} = \dfrac{6.025025 - 6}{0.005} = 5.005$.

   (b) The slope of M overestimates the slope of L because the graph of f is concave up near $x = 2$.

   (c) The slope of M is a better estimate of $f'(2)$ than is the slope of L' because the distance between $x = 2$ and $x = 2.005$ is less than the distance between $x = 2$ and $x = 2.01$. In other words, the slope of M is the average rate of change of f over a shorter x-interval (i.e., the x-interval is more nearly "instantaneous").

5. As the viewing window zooms in on the point $(2, 6)$, the graphs of f and L become indistinguishable.

7. $f'(-1) = \lim\limits_{h \to 0} \dfrac{f(-1+h) - f(-1)}{h}$

   $= \lim\limits_{h \to 0} \dfrac{((-1+h)^2 + (-1+h)) - 0}{h}$

   $= \lim\limits_{h \to 0} \dfrac{(1 - 2h + h^2) + (-1 + h)}{h}$

   $= \lim\limits_{h \to 0} \dfrac{-h + h^2}{h}$

   $= \lim\limits_{h \to 0} (-1 + h)$

   $= -1$

9. The difference quotient is the slope of the secant line through the points $(a, f(a))$ and $(a+h, f(a+h))$.

11. (a) $\dfrac{11.25 - 5.50}{6} \approx 0.96$ dollars/hr

    (b) $\dfrac{6.50 - 6.50}{4} = 0$ dollars/hr

    (c) The slope of the tangent line at 9:15 A.M. is approximately -4 dollars/hr.

13. $f'(3) \approx \dfrac{f(3.001) - f(3)}{0.001} = \dfrac{0.003001 - 0}{0.001} = 3.001$

15. $f'(0) \approx \dfrac{f(0.001) - f(0)}{0.001} \approx \dfrac{0.000999 - 0}{0.001} = 0.999$

17. $f'(1.99) \approx \dfrac{f(1.999) - f(1.99)}{1.999 - 1.99} \approx 102.89$

19. $f'(2.01) \approx \dfrac{f(2.05) - f(2.01)}{2.05 - 2.01} \approx 108$

21. $f'(1.9) \approx \dfrac{f(1.9) - f(1.97)}{1.9 - 1.97} \approx 4.357$

23. $f'(2) \approx \dfrac{f(2.02) - f(2)}{2.02 - 2} = 2.95$

25. $f(1.9) = 3.61$, $f(1.95) = 3.8025$, $f(2.0) = 4$, $f(2.01) = 4.0401$.

    (a) $f'(1.9) \approx \dfrac{f(1.95) - f(1.9)}{1.95 - 1.9} = 3.85$

SECTION 2.1 DEFINING THE DERIVATIVE

(b) $f'(2.0) \approx \dfrac{f(2.01) - f(2)}{2.01 - 2} = 4.01$

(c) $f'(2.01) \approx \dfrac{f(2) - f(2.01)}{2 - 2.01} = 4.01$

27. (a) $f'(1.9) \approx \dfrac{f(1.95) - f(1.9)}{1.95 - 1.9} \approx 6.86$

(b) $f'(2.0) \approx \dfrac{f(2.01) - f(2)}{2.01 - 2} \approx 7.43$

(c) $f'(2.01) \approx \dfrac{f(2) - f(2.01)}{2 - 2.01} \approx 7.43$

29. (a) This is the definition of the derivative of f at $x = 3$.

(b) $f(3 + h) = (3 + h)^2$ and $f(3) = 9$

(c) $(3 + h)^2 - 9 = (9 + 6h + h^2) - 9 = 6h + h^2$

(d) Since $h \neq 0$, $\dfrac{6h + h^2}{h} = 6 + h$

(e) $\lim_{h \to 0} 6 = 6$ and $\lim_{h \to 0} h = 0$, so $\lim_{h \to 0}(6 + h) = 6$

31. (a) This is the definition of the derivative of f at $x = 1$.

(b) $f(x) = x^2$ and $f(1) = 1$

(c) Since $x^2 - 1 = (x + 1)(x - 1)$, $(x^2 - 1)/(x - 1) = x + 1$ if $x \neq 1$.

(d) $\lim_{x \to 1} x = 1$ and $\lim_{x \to 1} 1 = 1$, so $\lim_{x \to 1}(x + 1) = 2$.

33. If $f(x) = \sqrt{x}$ and $a = 4$, then $f'(a) = \lim_{x \to a} \dfrac{f(x) - f(a)}{x - a} = \lim_{x \to 4} \dfrac{\sqrt{x} - \sqrt{4}}{x - 4} = \lim_{x \to 4} \dfrac{\sqrt{x} - 2}{x - 4} = \dfrac{1}{4}$.

35. If $f(x) = \cos x$ and $a = 0$, then
$f'(a) = \lim_{h \to 0} \dfrac{f(a + h) - f(a)}{h} = \lim_{h \to 0} \dfrac{\cos(0 + h) - \cos 0}{h} = \lim_{h \to 0} \dfrac{\cos h - 1}{h} = 0$.

37. The average rate of change of f over the interval $[1, 5]$ is $\dfrac{f(5) - f(1)}{5 - 1} = \dfrac{f(5) - 2}{4} = 3$. Therefore, $f(5) = 14$.

39. $f(1.998) = 1.4135063$; $f(1.999) = 1.4138600$; $f(2.000) = 1.4142136$; $f(2.001) = 1.4145671$; $f(2.002) = 1.4149205$.

(a) If the input is increased by 0.001, the output increases by about 0.000353.

(b) $f'(1.999) \approx \dfrac{f(2.000) - f(1.999)}{0.001} \approx 0.3536$; $f'(2) \approx 0.3535$; $f'(2.001) \approx 0.3534$

(c) It underestimates the exact value of $f'(2)$ because f is concave down near $x = 2$. (The increase in f over each interval of length 0.001 is decreasing.)

(d) $f''(2) \approx \dfrac{f'(2.001) - f'(2.000)}{0.001} \approx -0.100$

41. $f'(4) = \lim_{h \to 0} \dfrac{f(4 + h) - f(4)}{h} = \lim_{h \to 0} \dfrac{2(4 + h) + 1 - 9}{h} = \lim_{h \to 0} \dfrac{2h}{h} = \lim_{h \to 0} 2 = 2$

43. $f'(4) = \lim_{h \to 0} \dfrac{f(4 + h) - f(4)}{h} = \lim_{h \to 0} \dfrac{(4 + h)^3 - 64}{h} = \lim_{h \to 0} \dfrac{48h + 12h^2 + h^3}{h} = \lim_{h \to 0}(48 + 12h + h^2) = 48$

45. (a) No. A sketch of a function that is increasing and concave up over the interval [3, 5] makes it clear that the slope of the secant line through the points $(3, f(3))$ and $(5, f(5))$ is greater than $f'(3)$, the slope of the tangent line at the left endpoint of the interval.

More formally, since f is concave up on the interval [3, 5], f' is increasing over this interval. Therefore, $f(5) > f(3) + f'(3) \cdot (5-3) = f(3) + 2f'(3)$. This implies that
$$m = \frac{f(5) - f(3)}{5 - 3} > \frac{f(3) + 2f'(3) - f(3)}{2} = f'(3).$$

(b) No. A sketch of a function that is increasing and concave up over the interval [3, 5] makes it clear that the slope of the secant line through the points $(3, f(3))$ and $(5, f(5))$ is less than $f'(5)$, the slope of the tangent line at the right endpoint of the interval.

More formally, since f is concave up on the interval [3, 5], f' is increasing over this interval. Therefore, $f(3) > f(5) + f'(5) \cdot (3-5) = f(5) - 2f'(5)$. This implies that
$$m = \frac{f(5) - f(3)}{5 - 3} < \frac{f(5) - (f(5) - 2f'(5))}{2} = f'(5).$$

47. Since g is concave down over the interval [5, 8], the instantaneous rate of change of g at the left endpoint of the interval (i.e., $g'(5)$) is greater than the average rate of change of g over the entire interval. (This conclusion does not depend on the assumption that g is an increasing function on the interval.)

More formally, since g is concave down over the interval [5, 8], g' is a decreasing function on this interval. It follows that $g(8) < g(5) + g'(5) \cdot (8-5) = g(5) + 3g'(5)$. Thus,
$$a = \frac{g(8) - g(5)}{8 - 5} < \frac{g(5) + 3g'(5) - g(5)}{3} = g'(5).$$

49. $\lim\limits_{t \to 2} \dfrac{h(t) - h(2)}{t - 2} = h'(2) = 3$

51. $\lim\limits_{\Delta x \to 0} \dfrac{h(4 + \Delta x) - 5}{\Delta x} = h'(4) = -1$

53. $\lim\limits_{x \to 3} \dfrac{g(x) - g(3)}{x - 3} = g'(3) = 2$

55. The average rate of change of h over the interval $[-3, 5]$ is the slope of the secant line through the points $(-3, h(-3))$ and $(5, h(5))$. Since the average rate of change of h over the interval $[-3, 5]$ is 4, the (secant) line through the points $(-3, h(-3))$ and $(5, h(5))$ has slope 4. Since the line $y = 4x - 4$ has slope 4, it is possible that it is an equation of the secant line.
[NOTE: Not enough information about h is given to determine whether $y = 4x - 4$ is actually an equation of the secant line.]

57. The average rate of change of f over the interval [2, 6] is the slope of the secant line through the points $(2, f(2))$ and $(6, f(6))$. Thus, the line through these two points has slope 5. Since $f(2) = 3$, we can conclude that $y = 5x - 7$ is an equation of the secant line through the points $(2, f(2))$ and $(6, f(6))$.

SECTION 2.2 DERIVATIVES OF POWER FUNCTIONS AND POLYNOMIALS

§2.2 Derivatives of Power Functions and Polynomials

1. Division by zero is not a well-defined operation.

3. $\dfrac{df}{dx} = \sin(2x)$

5. $f'(-1) = 3$

7. $k'(x) = \lim\limits_{h \to 0} \dfrac{k(x+h) - k(x)}{h} = \lim\limits_{h \to 0} \dfrac{3-3}{h} = \lim\limits_{h \to 0} 0 = 0$

9. $l'(x) = 1 \implies l''(x) = 0$. Thus, $l''(1) = 0$. $q'(x) = 2x \implies q''(x) = 2$. Thus, $q''(2) = 2$. $c'(x) = 3x^2 \implies c''(x) = 6x$. Thus, $c''(3) = 18$.

11. Using the result derived in the Example, $f'(w) = \frac{1}{2}w^{-1/2}$. Thus, $f'(9) = \frac{1}{2}\dfrac{1}{\sqrt{9}} = \frac{1}{6}$.

13. (a)
$$x^k = x = x^1 \implies k = 1$$
$$x^k = x^2 \implies k = 2$$
$$x^k = \sqrt{x} = x^{1/2} \implies k = 1/2$$
$$x^k = 1/\sqrt{x} = x^{-1/2} \implies k = -1/2$$
$$x^k = \dfrac{x^5}{\sqrt{x^3}} = \dfrac{x^5}{x^{3/2}} = x^{7/2} \implies k = 7/2$$
$$x^k = x^\pi \implies k = \pi$$

(b) $\dfrac{d}{dx} x = 1$

$\dfrac{d}{dx} x^2 = 2x$

$\dfrac{d}{dx} \sqrt{x} = \frac{1}{2}x^{-1/2} = \dfrac{1}{2\sqrt{x}}$

$\dfrac{d}{dx}(1/\sqrt{x}) = -\frac{1}{2}x^{-3/2} = -\dfrac{1}{2\sqrt{x^3}}$

$\dfrac{d}{dx}(x^5/\sqrt[3]{x}) = \frac{7}{2}x^{5/2}$

$\dfrac{d}{dx} x^\pi = \pi x^{\pi - 1}$

15. $s'(x) = \frac{1}{2}x^{-1/2}$ so $s''(x) = -\frac{1}{4}x^{-3/2}$.

17. $f'(x) = 3 \cdot 2x + 0 = 6x$

19. $f'(x) = 4 \cdot 5x^4 + 3 \cdot 2x - 1 = 20x^4 + 6x - 1$

21. $f(x) = x^3 - \sqrt[3]{x} = x^3 - x^{1/3}$. Therefore, $f'(x) = 3x^2 - \frac{1}{3}x^{-2/3}$

23. $f(x) = 4\sqrt{x^3} - 5x^{-2/3} = 4x^{3/2} - 5x^{-2/3}$. Therefore,
$f'(x) = 4 \cdot \dfrac{3}{2}x^{1/2} - 5 \cdot (-2/3)x^{-5/3} = 6\sqrt{x} + \dfrac{10}{3}x^{-5/3}$.

25. $f''(x) = 6$

27. $f''(x) = 80x^3 + 6$

29. $f''(x) = 6x + \frac{2}{9}x^{-5/3}$

31. $f''(x) = 3x^{-1/2} - \frac{50}{9}x^{-8/3}$

33. Since $\frac{d}{du}u^3 = 3u^2$, $f'(x) = 3(x+2)^2$.

35. Since $\frac{d}{du}u^{-1} = -u^{-2}$, $f'(x) = -(x+4)^{-2}$.

37. Let $f(x) = x^3 - 4$. Since $f'(x) = 3x^2 - 4$, $f'(2) = 8$. Thus, the tangent line has slope 8. Since $f(2) = 0$, the tangent line is described by the equation $y = f'(2)(x-2) + f(2) = 8x - 16$.

39. The tangent line is described by the equation $y = f'(1)(x-1) + f(1) = 4(x-1) + 5 = 4x + 1$.

41. The tangent line is described by the equation $y = h'(1)(x-1) + h(1) = 4(x-1) + 7 = 4x + 3$.

43. The tangent line is described by the equation $y = l'(1)(x-1) + l(1) = 8(x-1) + 10 = 8x + 2$.

45. $f(x) = (x+2)(3x-4) = 3x^2 + 2x - 8 \implies f'(x) = 6x + 2$

47. $f(x) = \dfrac{x^3 - 4x + 1}{\sqrt{x}} = x^{5/2} - 4x^{1/2} + x^{-1/2} \implies f'(x) = \dfrac{5}{2}x^{3/2} - 2x^{-1/2} - \dfrac{1}{2}x^{-3/2}$

49. Let $f(x) = x^3 - 6x^2$. Then, $f'(x) = 3x^2 - 12x$ and $f''(x) = 6x - 12$. Thus, the curve has a point of inflection at $x = 2$. The tangent line at $x = 2$ has slope $f'(2) = -12$ and passes through the point $(2, -16)$; an equation of this tangent line is $y = -12(x-2) - 16 = -12x + 8$.

51. The volume of a sphere of radius r is $V(r) = \frac{4}{3}\pi r^3$. Thus, $V'(r) = 4\pi r^2$.

53. (a) $\dfrac{dP}{dV} = -\dfrac{nRT}{(V-nb)^2} + \dfrac{2n^2 a}{V^3}$

    (b) Yes, because $\dfrac{dP}{dV} < 0$.

55. Let $f(x) = (1+x)^r$ and $g(x) = 1 + rx$. Then, $f(0) = g(0)$, $f'(x) = r(1+x)^{r-1}$, and $g'(x) = r$. Since $r \geq 1$, $(1+x)^{r-1} \geq 1$ for all $x \geq 0$, it follows that $f'(x) \geq r = g'(x)$ for all $x \geq 0$. Therefore, by the speed limit law, $f(x) \geq g(x)$ if $x \geq 0$.

57. First, note that $\lim\limits_{w \to 1} \dfrac{g(w) - g(1)}{w} = g'(1)$. Now, $g(w) = 2/w + 3w^4 = 2w^{-1} + 3w^4$ so $g'(w) = -2w^{-2} + 12w^3$ and, therefore, $g'(1) = 10$. Thus, $\lim\limits_{w \to 1} \dfrac{g(w) - g(1)}{w} = 10$.

59. If $n \geq 1$, $f^{(n)}(x) = (-1)^n \dfrac{(2n-3)(2n-5)\cdots(3)(1)(-1)}{2^n} x^{-(2n-1)/2}$.

61. Let $f(x) = 2x^6 + 9x^5 + 10x^4 - x + 2$. Then,
$f''(x) = 60x^4 + 180x^3 + 120x^2 = 60x^2(x^2 + 3x + 2) = 60x^2(x+1)(x+2)$ so f has two points of inflection: $x = -2$ and $x = -1$. [NOTE: $x = 0$ is *not* a point of inflection since f'' does not change sign at $x = 0$.]

63. Since $f(x)$ is not defined if $x < 0$, the (two-sided) limit that defines $f'(0)$ cannot be evaluated.

65. Yes. Every polynomial can be written in the form $p(x) = a_0 + a_1 x + a_2 x^2 + \cdots + a_{n-1}x^{n-1} + a_n x^n$, where $a_0, a_1, \ldots, a_n$, are constants and n is a positive integer. Since $p'(x) = a_1 + 2a_2 x + \cdots + (n-1)a_{n-1}x^{n-2} + na_n x^{n-1}$ also has this form, it is also a polynomial. (The derivative of an nth-degree polynomial is a polynomial of degree $n-1$.)

67. $\dfrac{d^n}{dx^n} p(x) = n!$

(a) $\dfrac{d^n}{dx^n} p(x) = n!$

(b) $\dfrac{d^k}{dx^k} x^m$ is zero if $m < k$, $k!$ if $m = k$, and $m(m-1)\cdots(m-k+1)x^{m-k}$ if $m > k$. Thus, $p^{(k)}(0) = k! a_k$

69. $f'(x) = \lim\limits_{h\to 0} \dfrac{f(x+h) - f(x)}{h}$

$= \lim\limits_{h\to 0} \dfrac{1/(x+h+5) - 1/(x+5)}{h}$

$= \lim\limits_{h\to 0} \dfrac{(x+5) - (x+h+5)}{h(x+h+5)(x+5)}$

$= \lim\limits_{h\to 0} \dfrac{-1}{(x+h+5)(x+5)}$

$= -\dfrac{1}{(x+5)^2} = -(x+5)^{-2}$

71. $f'(x) = \lim\limits_{h\to 0} \dfrac{f(x+h) - f(x)}{h}$

$= \lim\limits_{h\to 0} \dfrac{(x+h+2)^3 - (x+2)^3}{h}$

$= \lim\limits_{h\to 0} \dfrac{3h(x+2)^2 + 3h^2(x+2) + h^3}{h}$

$= \lim\limits_{h\to 0} \left(3(x+2)^2 + 3h(x+2) + h^2\right) = 3(x+2)^2$

§2.3 Limits

1. (a) $\lim_{x \to 3} f(x) = \lim_{x \to 3} x^2 = 9$

    (b) No, because $f(3) = 0$.

3. (a) $\dfrac{x^2 - 9}{x + 3} = \dfrac{(x+3)(x-3)}{x+3}$. Therefore, if $x \neq -3$, then $x + 3 \neq 0$ so $\dfrac{(x+3)(x-3)}{x+3} = x - 3$.

    (b) $\lim_{x \to -3} \dfrac{x^2 - 9}{x + 3} = \lim_{x \to -3} (x - 3) = -6$

    (c) $f'(-3) = \lim_{x \to -3} \dfrac{x^2 - 9}{x + 3} = -6$

5. $g'(0) = \lim_{h \to 0} \dfrac{g(h) - g(0)}{h} = \lim_{h \to 0} \dfrac{h^{1/5}}{h} = \lim_{h \to 0} \dfrac{1}{h^{4/5}} = \infty$

7. First, note that $f'(x) = \begin{cases} 1 & x > 0 \\ -1 & x < 0. \end{cases}$

    (a) $\lim_{x \to 0^+} f'(x) = \lim_{x \to 0^+} 1 = 1$

    (b) $\lim_{x \to 0^-} f'(x) = \lim_{x \to 0^-} -1 = -1$

    (c) $\lim_{x \to 0} f'(x)$ does not exist because $\lim_{x \to 0^+} f'(x) \neq \lim_{x \to 0^-} f'(x)$.

9. (a) To sketch the graph, first draw the horizontal line joining $(0, 2)$ and $(5, 2)$. Then, erase the six points with x-coordinates $0, 1, 2, \ldots, 5$. Finally, blacken the six points $(0, 1), (1, 1), (2, 1), \ldots, (5, 1)$.

    (b) $\lim_{x \to 4} f(x) = 2$ because $f(x) = 2$ for all nearby values of $x \neq 4$.

11. $\lim_{x \to a} f(x) = 2$ for *all* values of a. (The point of the problem is that, in this case, taking the limit "forgives" the fact that the function has funny values at integer inputs.)

13. $\lim_{x \to 0} f(x) = 2$

15. $\lim_{x \to -1^-} f(x) = 1$, and $\lim_{x \to -1^+} f(x) = 2$. Therefore, the two-sided limit $\lim_{x \to -1}$ does not exist.

17. $\lim_{x \to 2^+} f(x) = 1$

19. $\lim_{x \to -2^-} f(x) = 0$

21. $\lim_{x \to -4^+} f(x) = -2$

23. $\lim_{x \to 3} f'(x) = -1$ since f is a linear function with slope -1 for all values of x "near" $x = 3$.

25. $\lim_{x \to -1^-} f'(x) = 1$ since f is a linear function with slope 1 for all nearby values of x to the left of $x = -1$.

27. $\lim_{x \to -4^+} g(x) = 0$

29. $\lim_{x \to -2} g(x) = 3/2$

31. $\lim_{x \to -1^-} g(x) = 2$, and $\lim_{x \to -1^+} g(x) = 1$. Therefore, the two-sided limit $\lim_{x \to -1} g(x)$ does not exist.

SECTION 2.3 LIMITS

33. $\lim\limits_{x \to 1} g(x) = -2$

35. $\lim\limits_{x \to 3} g(x) = 1$

37. (a) The secant line passes through the points $(3, 12)$ and $(t, t^2 + t)$. Therefore, for any $t \neq 3$ its slope is
$$s(t) = \frac{(t^2 + t) - 12}{t - 3} = \frac{(t - 3)(t + 4)}{t - 3} = t + 4.$$

(b) $\lim\limits_{t \to 3} s(t) = \lim\limits_{t \to 3}(t + 4) = 7.$

(c) The answer to part (b) means that at $x = 3$ the line tangent to the f-graph has slope 7. The tangent line passes through $(3, 12)$ with slope 7, so it is described by the equation $y = 7(x - 3) + 12 = 7x - 9$. A plot of this line and f over the interval $[2, 4]$ illustrates this.

39. $\lim\limits_{x \to 3} f(x) = \lim\limits_{x \to 3} 3x - 5 = 4$

41. $\lim\limits_{x \to 2^-} f(x) = \lim\limits_{x \to 2^-} 3 - x^2/2 = 1$ and $\lim\limits_{x \to 2^+} f(x) = \lim\limits_{x \to 2^+} 3x - 5 = 1$. Therefore, $\lim\limits_{x \to 2} f(x) = 1$.

43. $\lim\limits_{x \to 1}(f(x) + g(x)) = \lim\limits_{x \to 1} f(x) + \lim\limits_{x \to 1} g(x) = 5 + (-2) = 3$

45. $\lim\limits_{x \to 1} f(x) \cdot g(x) = \left(\lim\limits_{x \to 1} f(x)\right) \cdot \left(\lim\limits_{x \to 1} g(x)\right) = 5 \cdot -2 = -10$

47. By definition of continuity at $x = 3$, $\lim\limits_{x \to 3} f(x) = f(3) = -1$.

49. $\lim\limits_{x \to 0}(f(x) + g(x)) = \lim\limits_{x \to 0} f(x) + \lim\limits_{x \to 0} g(x) = 2 + 0 = 2$

51. $\lim\limits_{x \to 1} f(x) \cdot g(x)$ does not exist because $\lim\limits_{x \to 1} f(x)$ does not exist (the left and right limits are not equal)

53. $\lim\limits_{x \to 0} \dfrac{f(x)}{g(x)}$ does not exist because $\lim\limits_{x \to 0^-} \dfrac{f(x)}{g(x)} = \infty$ and $\lim\limits_{x \to 0^+} \dfrac{f(x)}{g(x)} = -\infty$.

55. $\lim\limits_{x \to -2} x^2 g(x) = \left(\lim\limits_{x \to -2} x^2\right) \cdot \left(\lim\limits_{x \to -2} g(x)\right) = 4 \cdot 1 = 4$

57. $\lim\limits_{x \to 0} \dfrac{f(x) - 2}{x} = f'(0) = 0$

59. $\lim\limits_{h \to 0} \dfrac{g(-2 + h) - 3/2}{h} = g'(-2) = 1/2$

61. Yes, because the existence of the limit depends only on how f behaves *near* $x = 2$.

63. **Must** be true. Continuity of f at $x = 3$ means that $\lim\limits_{x \to 3} f(x) = f(3)$, so $f(3) = 17$; in particular, $x = 3$ is in the domain of f.

65. **Must** be true. Since $\lim\limits_{x \to 3} f(x) = 17$ we must have $\lim\limits_{x \to 3^-} f(x) = \lim\limits_{x \to 3^+} f(x) = 17$.

67. (a) To assure continuity at $x = 2$ the two parts of the definition must "agree" at $x = 2$. Therefore, since $f(2) = 3 \cdot 2 - 5 = 1$, $1 = \lim\limits_{x \to 2^-} f(x) = \lim\limits_{x \to 2^-}(ax^2 + 3) = a(2)^2 + 3 = 4a + 3$ must hold. Thus, $4a + 3 = 1$ so $a = -1/2$.

(b) No, the graph of f has a "kink" at $x = 2$. More formally, f is differentiable at $x = 2$ if $\lim\limits_{x \to 2} \dfrac{f(x) - f(2)}{x - 2}$ exists and is a finite number. However,

$$\lim_{x \to 2^-} \frac{f(x) - f(2)}{x - 2} = \lim_{x \to 2^-} \frac{(ax^2 + 3) - (4a + 3)}{x - 2} = \lim_{x \to 2^-} \frac{a(x^2 - 4)}{x - 2} = \lim_{x \to 2^-} 4a = -2$$

(since $a = -1/2$) and

$$\lim_{x \to 2^+} \frac{f(x) - f(2)}{x - 2} = \lim_{x \to 2^+} \frac{(3x - 5) - 1}{x - 2} = \lim_{x \to 2^+} \frac{3x - 6}{x - 2} = \lim_{x \to 2^+} 3 = 3.$$

Therefore, since the limit $\lim\limits_{x \to 2} \dfrac{f(x) - f(2)}{x - 2}$ does not exist, $f'(2)$ is undefined and f is not differentiable at $x = 2$.

SECTION 2.4 USING DERIVATIVE AND ANTIDERIVATIVE FORMULAS

§2.4 Using derivative and antiderivative formulas

1. (a) No. The minimum value of f on I can occur only at a stationary point of f that is in I or at an endpoint of I.

   (b) Yes. The maximum value of f on I could occur at an endpoint of I.

3. Since F is an antiderivative of f, $F'(4) = f(4) = 5$.

5. Since F is an antiderivative of f, $F'' = f'$. Thus, F has an inflection point wherever f' changes sign. Since f has local extrema at $x = 1.14$ and at $x = 8.19$, f' changes sign at these points. Therefore, these points are inflection points of F.

7. $F''(5) = f'(5) \approx -1.5$.

9. $F(0) = 10$ and $F'(0) = f(0) = 7$ so L is the line with slope 7 that passes through the point $(0, 10)$. Thus, L is described by the equation $y = 7x + 10$.

11. No. $F'(x) = f(x) > 0$ over the interval $[0, 5]$ implies that F is increasing over this interval. Since $F(0) = 10$, this means that $F(5) > F(0) = 10$.

13. Since $f = F'$, f must be positive on the intervals where F is increasing. Thus, f is positive on the intervals $(3.2, 6)$ and $(13.3, 15)$.

15. Using the fact that $f(x) = F'(x)$ and the graph of F, it is clear that $-2 < f(9) < 0$, that $f(3) \approx 0$, and that $0 < f(6) < f(14)$. Thus, $-2 < f(9) < f(3) < f(6) < f(14)$.

17. Since $f' = F''$, f is increasing (i.e., $f' > 0$) on the intervals where F is concave up. Thus, f is increasing on the intervals $(0, 4.5)$ and $(10.5, 15)$.

19. $f'(x) = 3x^2 - 6x = 3x(x - 2)$

    (a) f is increasing on the intervals where $f' > 0$. Thus, f is increasing on the intervals $(-\infty, 0)$ and $(2, \infty)$.

    (b) f is decreasing on the intervals where $f' < 0$. Thus, f is decreasing on the interval $(0, 2)$.

    (c) f has a local maximum at any point where the sign of f' changes from positive to negative. Thus, f has a local maximum only at $x = 0$.

    (d) f has a local minimum at any point where the sign of f' changes from negative to positive. Thus, f has a local minimum only at $x = 2$.

21. $g'(w) = 63w^8 - 126w^6 = 63w^6(w^2 - 2)$

    (a) g is increasing on the intervals where $g' > 0$. Thus, g is increasing on the intervals $(-\infty, -\sqrt{2})$ and $(\sqrt{2}, \infty)$.

    (b) g is decreasing on the intervals where $g' < 0$. Thus, g is decreasing on the interval $(-\sqrt{2}, \sqrt{2})$.

    (c) g has a local maximum at any point where the sign of g' changes from positive to negative. Thus, g has a local maximum only at $x = -\sqrt{2}$.

    (d) g has a local minimum at any point where the sign of g' changes from negative to positive. Thus, g has a local minimum only at $x = \sqrt{2}$.

23. $f'(x) = 3x^2 - 3 = 3(x^2 - 1) = 3(x - 1)(x + 1)$. Thus, f is increasing on the intervals $(-\infty, -1)$ and $(1, \infty)$; f is decreasing on $(-1, 1)$. Therefore, the minimum value of f over the interval $[0, 2]$ is $f(1) = 3$; the maximum value of f on the interval $[0, 2]$ is $f(2) = 7$.

25. $f'(x) = \frac{1}{2}x^{-1/2} + 3x^{-2}$. Since $f' > 0$ over the interval $[0.25, 1]$, the maximum value of f is $f(1) = -2$ and the minimum value of f is $f(0.25) = -11.5$.

27.
$$\left(\frac{1}{2}x^6 + \frac{7}{5}x^5 - \frac{1}{9}x^3 + 11x + 13\right)' = \frac{1}{2} \cdot 6x^5 + \frac{7}{5} \cdot 5x^4 - \frac{1}{9} \cdot 3x^2 + 11 = 3x^5 + 7x^4 - x^2/3 + 11 = p(x).$$

29. Since F is an antiderivative of f, $F'(x) = f(x) = x \sin x$.

31. No, because $G'(x) = 3x^2 \neq g(x)$.

33. Since $G'(x) = g(x)$, G is an antiderivative of g. This means that any function of the form $G(x) + C$, where C is a constant, is an antiderivative of g. Thus, for example, $x \sin(e^x) - 1$, $x \sin(e^x)$, and $x \sin(e^x) + 2$ are antiderivatives of g. (Infinitely many other correct answers are possible.)

35. (a) If H is an antiderivative of h, $H'(x) = 3x^2$. Since the derivative of x^k is kx^{k-1}, we expect that H will involve a constant multiple of x^3 (i.e., a term of the form Ax^3). Furthermore, since antiderivatives of a function may differ from each other by a constant, we must allow for the possibility of an additive constant. Thus, it is natural to guess that $H(x) = Ax^3 + B$, where A and B are constants.

(b) Assume $H(x) = Ax^3 + B$. Since $H(0) = 1$, $B = 1$ must be true. Since $H'(x) = 3Ax^2$, $A = 1$ must be true.

37. (a) Yes — $f'(0) = -1$. (The graph is "smooth" and locally linear at $x = 0$.)

(b) $f'(x) = \begin{cases} -1 & -1 \leq x \leq 0 \\ 4x - 1 & 0 < x \leq 1 \end{cases}$

39. Since $q'(x) = 2ax + b$, q has a stationary point (i.e., $q'(x) = 0$) only at the point $x = -b/2a$.

41. Every cubic polynomial can be written in the form $c(x) = Ax^3 + Bx^2 + Cx + D$, where A, B, C, and D are constants. Since $c''(x) = 6Ax + 2B$, c will have an inflection point at $x = 1$ only if $B = -3A$. Furthermore, the concavity of c will change from concave up to concave down at $x = 1$ only if the sign of c'' changes from positive to negative there. This implies that $A < 0$ must be true. Thus, any polynomial of the form $c(x) = Ax^3 - 3Ax^2 + Cx + D$, where A, C, and D are constants and $A < 0$, will have the desired property. An example is $c(x) = 3x^2 - x^3$.

43. Since $f'(x) = x(A + x)$, the stationary points of f are at $x = 0$ and $x = -A$. Therefore, the maximum value of f on I could occur at $x = -4, x = 3, x = 0$, or $x = -A$ (i.e., the endpoints of I and the stationary points of f).

45. Since $q'(x)$ is a polynomial of degree at most $n - 1$, q' has at most $n - 1$ roots. Thus, q has at most $n - 1$ stationary points. Since local extrema can occur only at stationary points, q has at most $n - 1$ local extrema.

47. No—the maximum value of $f(x) = 4x^3 - x^4$ is $f(3) = 27$. To see this, note that $f'(x) = 12x^2 - 4x^3 = 4x^2(3 - x)$ so f is increasing on $(-\infty, 3)$ and decreasing on $(3, \infty)$.

49. (a) If p has a local minimum at $(2, -9)$, $x = 2$ must be a stationary point of p. Since $p'(x) = 3x^2 + a$, $p'(2) = 0$ implies that $a = -12$. Now, since $p(2) = -9$, $b = 7$ must be true. Thus, $p(x) = x^3 - 12x + 7$.
[NOTE: Since $p''(x) = 6x$, $p''(2) = 12 > 0$ so the point $(2, -9)$ is a local minimum of p.]

51. (a) Let $q(x) = ax^3 + bx^2 + cx + d$. Then, $q'(x) = 3ax^2 + 2bx + c$, and $q''(x) = 6ax + 2b$. From the information given in the problem, we know that $q(0) = 2 = d$, $q(5) = 0 = 125a + 25b + 5c + d$, $q'(0) = 0 = c$, $q'(5) = 0 = 75a + 10b$, $q''(0) \leq 0$, and $q''(5) \geq 0$. Solving these four equations for the unknowns a, b, c, and d, we find $a = 4/125$, $b = -6/25$, $c = 0$, and $d = 2$. Thus, $q(x) = \frac{4}{125}x^3 - \frac{6}{25}x^2 + 2$.

53. No such value of k exists. To see this, let $f(x) = (8x + k)/x^2 = 8/x + k/x^2$. Then, $f'(x) = -8/x^2 - 2k/x^3$ and $f'(4) = 0 \implies k = -16$. However, with this value of k, $f''(4) < 0$ which implies that f has a local *maximum* at $x = 4$.

SECTION 2.4 USING DERIVATIVE AND ANTIDERIVATIVE FORMULAS

55. Since $v(t)$ is the rate of change (or derivative) of the position function $s(t)$, $s(t) = t^3/3 + C$ for some constant C. Thus, the distance the particle travels between $t = 1$ and $t = 3$ is $s(3) - s(1) = 26/3$.

57. The distance from the origin to a point (x, y) on the curve $xy = 4$ is $d(x) = \sqrt{x^2 + 16/x^2}$. Since the distance function is nonnegative, the function $f(x) = \bigl(d(x)\bigr)^2 = x^2 + 16/x^2$ will achieve its minimum value at the same x-value as $d(x)$ does. Now, $f'(x) = 2x - 32/x^3$ so f has stationary points at $x = \pm 2$. Since $f''(\pm 2) = 8$, both of these stationary points correspond to local minima of f. Thus, the minimum distance is $d(2) = 2\sqrt{2}$.

59. Let's use the following notation: $v =$ speed; $fc(v) =$ hourly fuel cost at speed v; $tc(v) =$ total cost for entire trip, at speed v. The problem, then, is to minimize tc; in theory, *any* positive value of v is possible.

    By the first statement, $fc = kv^2$, for some constant v. The second statement says that $fc(10) = 3000 = k \cdot 100$ if $v = 10$, i.e., that $k = 30$. Thus $fc(v) = 30v^2$. Adding fixed costs says that the total hourly cost of operation, at speed v, is $30v^2 + 12000$.

    Notice next that a trip of D km takes D/v hours. Therefore the total cost of such a trip is given by

    $$tc(v) = \text{cost per hour} \times \text{number of hours} = (30v^2 + 12000)(D/v) = D \cdot (30v + 12000/v).$$

    A look at the graph of tc shows that it has one local minimum—the one we want. To find it, we look for roots of tc':

    $$tc'(v) = D \cdot (30 - 12000/v^2) = 0 \iff v = 20.$$

    Conclusion: 20 km/hr is the optimum speed—regardless of D.

61. If the box has width x, length y, and depth z then, by assumption, $y = 2x$ and $xyz = 36{,}000$. (Therefore, $2x^2 z = 36{,}000$, so $z = 18{,}000/x^2$.) The total area—which we want to minimize—is $A = xy + 2xz + 2yz$. (Draw a picture.)

    Although A starts life as a function of 3 variables, with the equations above we can reduce everything to one variable, say x. Here goes:

    $$A = xy + 2xz + 2yz = x(2x) + 2x(18000/x^2) + 2(2x)(18000/x^2) = 2x^2 + 108000/x.$$

    Thus $A'(x) = 4x - 108000/x^2 = 0 \iff x^3 = 27000 \iff x = 30$. Thus the optimal box has dimensions $x = 30$, $y = 60$, $z = 20$—all in inches. The area of such a box is $A(30) = 2 \cdot 30^2 + 108000/30 = 5400$ square inches, or $5400/144 = 37.5$ square feet; at these prices, it costs \$3.75 to build.

63. In standard form, the equation of the line $2x + 3y = a$ is $y = -\frac{2}{3}x + \frac{1}{3}a$ so the line has slope $-2/3$. Therefore, $f'(3) = -2/3$.

    The slope of the line tangent to f at x is $f'(x) = 2bx$, so we have $f'(3) = 6b = -2/3$ and, therefore, $b = -1/9$.

    Now, $f(3) = -\frac{1}{9} \cdot 3^2 = -1$ so the line $2x + 3y = a$ must pass through the point $(3, -1)$. This implies that $a = 3$.

65. The curve $y = x^2 + c$ has slope 4 only at $x = 2$. Thus, $x = 2$ must be the point of tangency. From this it follows that $c = 7$.

67. $F(x) = \ln x$ is an antiderivative of $f(x) = 1/x$ for $x > 0$. Since f is an odd function, $F(-x) = \ln(-x)$ is an antiderivative of f for $x < 0$.

§2.5 Differential Equations; Modeling Motion

1. No. The function $f(x) = 3x^2 + 5x + 6$ is a solution of the DE $f' = 6x + 5$, but there are infinitely many other solutions. Every function of the form $g(x) = 3x^2 + 5x + C$, where C is a constant, is a solution of the DE. Therefore, f is not the unique solution of the DE.

3. (a) $y(x) = 3x$ is one solution. In general, $y(x) = 3x + C$, where C is a constant, is a solution of the DE.

   (b) There are infinitely many solutions of the DE. They differ from each other by a constant.

5. Yes. $g(x) = \sqrt{x+4} = (x+4)^{1/2} \implies g'(x) = \frac{1}{2}(x+4)^{-1/2} = \frac{1}{2g(x)}$.

7. $y' = y$ is an example of a first-order DE.

   $y'' = -32$ is an example of a second-order DE.

9. Solutions of the DE $y' = 6x + 5$ have the form $y(x) = 3x^2 + 5x + C$, where C is a constant. The initial condition $y(1) = 2$ means that the function of this form with $C = -6$ is to the unique solution of the IVP.

11. Functions of the form $y(t) = \pm\sqrt{t+C}$ are solutions of the DE $y' = 1/2y$. Only the solution $y(t) = -\sqrt{t+3}$ satisfies the initial condition $y(1) = -2$.

13. The minus sign means that the acceleration due to gravity is downwards, towards the earth.

15. $h(t) = -4.9t^2 + 98t + 3$ and $v(t) = -9.8t + 98$

    (a) The cannonball is highest at $t = 10$ (i.e., when $v(t) = 0$). Since $h(10) = 493$, the cannonball rises to 493 meters.

    (b) $h(t) = 0$ when $t = 10 + \sqrt{4930}/7 \approx 20.03$. Thus, the cannonball remains airborne for about 20.03 seconds.

17. $y' = 3x^2$ so $xy' = x \cdot 3x^2 = 3x^3 = 3y$

19. Yes, $y(t) = t^2/2 \implies y' = t$

21. No. $y' - 2y/t = 2t^3 + 3t - 2(t^4/2 + 3t^2/2 + 1/4)/t = t^3 - 1/(2t) \neq t^3$.

23. Let $y(t) = \sqrt[3]{x+C} = (x+C)^{1/3}$. Then, $y'(t) = \frac{1}{3}(x+C)^{-2/3} = \frac{1}{3y^2}$.

25. No, because $y'' - 4xy' + 4y = 2 - 4x(2x) + 4x^2 = 2 - 4x^2 \neq 0$.

27. $y' = -1/x^2$ and $y'' = 2/x^3$ so $x^3y'' + x^2y' - xy = 2 - 1 - x(x^{-1} - 1) = x$.

29. At time $t = 2$ seconds, the object is falling at a rate of 3 meters per second.

31. The object's acceleration is downward and proportional to to the square of its velocity.

33. $p'(t) = v(t) = +15$. (The sign is positive because the object is moving eastward, the positive direction.)

35. $p'(t) = v(t) = k\sqrt{p}$

37. The solution of the IVP $h'(t) = -3$ and $h(0) = 100$ is $h(t) = 100 - 3t$. Thus, $h(10) = 70$. In words, at time $t = 10$ seconds, the object is 70 meters above ground level.

39. Let $s(t)$ denote the position of the object at time t. Since the object's acceleration is -8 meters per second per second, $s''(t) = -8$. From this it follows that $s'(t) = -8t + 14$ (since the object's velocity at time $t = 0$ is 14 meters per second) and that $s(t) = -4t^2 + 14t + s_0$, where s_0 is the position of the object at time $t = 0$. Therefore, the distance traveled by the object between $t = 0$ and $t = 3$ is $s(3) - s(0) = 6$ meters.

SECTION 2.5 DIFFERENTIAL EQUATIONS; MODELING MOTION

41. (a) The DE $P' = 0.08P$ means that the magnitude of the instantaneous rate of growth of the amount on deposit is 8% of the amount on deposit — exactly what is meant by an continuously compounded interest rate of 8

    (b) The DE $P'(t) = 0.1P(t)$ models a continously compounded growth rate of 10% per year. The DE $P'(t) = -100$ models a account balance that is declining at the constant rate of $100 per year. Thus, the DE $P'(t) = 0.1P(t) - 100$ expresses Second National Bank's policy.

43. The difference between the temperature of the object and its surrounds is $T - S$. Thus, the physical law is described by the DE $T' = k(T - S)$, where k is a constant.

45. If P is "small" (i.e., low skill level), then $M - P$ is "large" (i.e., there is a lot of room for improvement). Since k is a positive constant, this implies that P' is "large" and positive (i.e., performance is improving rapidly). If P is close to its maximum value, $M - P$ is "small" so P' is also "small" (i.e., performance is improving slowly).

47. (a) If $P(0) = 1000$, then $P' = 0$ so the model predicts that the size of the rabbit population remains constant (i.e., births = deaths).

    (b) If $P(0) = 1500$, then $P' < 0$ so the model predicts that the size of the rabbit population decreases.

    (c) If $P(0) = 250$, then $P' > 0$ so the model predicts that the size of the rabbit population increases.

    (d) The rate of change of the rabbit population is maximal at the value of P where $\frac{d}{dP}P'$ changes sign from positive to negative. Since

    $$\frac{d}{dP}P' = \frac{d}{dP}(kP(1000 - P)) = \frac{d}{dP}(1000kP - kP^2) = 1000k - 2kP = 2k(500 - P),$$

    the maximal rate of change occurs at $P = 500$ (i.e., at one-half the maximum population).

49. (a) It sounds more or less reasonable. As the goal approaches, people's ardor to contribute seems likely to cool somewhat. On the other hand, as the goal becomes really near, some people might give money to get it over with.

    (b) As a DE, Neuman's law of cooling of enthusiasm says $y' = k(y - 65)$.

51. (a) $y = x^3 \implies y' = 3x^2 = 3(x^3)^{2/3} = 3y^{2/3}$. Furthermore, $x = 0 \implies y = x^3 = 0$. Thus, $y = x^3$ is a solution of the IVP $y' = 3y^{2/3}$, $y(0) = 0$.

    (b) $f'(x) = \begin{cases} 3(x+1)^2 & x \le -1 \\ 0 & -1 < x < 2 \\ 3(x-2)^2 & x \ge 2. \end{cases}$

    As in part (a), $f'(x) = 3(f(x))^{2/3}$. Since $f(0) = 0$, f is also a solution of the given IVP.
    [NOTE: The point here is that not every IVP has a *unique* solution.]

53. (a) If the point $P = (x, y)$ is on the curve, then the endpoints of the line segment RS are $(0, 2y)$ and $(2x, 0)$. Furthermore, the slope of the line tangent to the curve at P is the slope of the line segment RS (i.e., $-y/x$). Therefore, at each point on the curve, $y' = -y/x$.

    (b) If $y = 1/x = x^{-1}$, then $y' = -x^{-2} = -x^{-1}/x = -y/x$.

§2.6 Derivatives of Exponential and Logarithmic Functions; Modeling Growth

1. The number e is the base for which the curve $y = b^x$ has slope 1 at $x = 0$.

3. Yes — $(y(t))' = (2e^t)' = 2e^t = y(t)$.

5. The general solution of the DE $y' = y$ is $y(x) = Ae^x$. The initial condition $y(1) = 2e = Ae^1 \implies A = 2$. Thus, $y(x) = 2e^x$ is the unique solution of the IVP.

7. $h'(b) = 1/c$. Because g and h are inverse functions, the slope of the g-graph at the point (a, b) is the reciprocal of the slope of the h-graph at the point (b, a).

9. Antiderivatives of h have the form $H(x) = 5^x/\ln 5 + C$, where C is a constant. Thus, for example, $5^x/\ln 5$ and $5^x/\ln 5 + 1$ are both antiderivatives of h.

11. Antidifferentiating each term on the right side of the DE leads to the solution $y(x) = 2e^x - 2^x/\ln 2 + 2x$.

13. (a) The general solution of the DE $y' = 3y$ is $y(x) = Ae^{3x}$. The initial condition implies that $A = 37$, so the solution of the IVP is $y(x) = 37e^{3x}$.

    (b) One; the solution is unique.

15. $f'(x) = 2e^x$ (since π is a constant, its derivative is zero)

17. $f'(x) = (\ln 2)2^x + 2x$

19. $f'(x) = -2/x$

21. $f''(x) = 2e^x$

23. $f''(x) = (\ln 2)^2 2^x + 2$

25. $f''(x) = 2x^{-2} = 2/x^2$

27. $F(x) = 2e^x + \pi x + C$

29. $F(x) = 2^x/\ln 2 + x^3/3 + 2x + C$

31. Let $f(x) = e^x$. Then $f(x) = f'(x) = e^x$ and $f(0) = f'(0) = 1$. Thus, the line tangent to the curve $y = e^x$ at $x = 0$ has slope 1 and passes through the point $(0, 1)$; it is described by the equation $y = x + 1$.

33. The slope of the curve $y = f(x)$ at the point (x, y) is $y' = f'(x)$. Since $f(x) = 3^x \implies f'(x) = (\ln 3)3^x$, the slope of the curve $y = 3^x$ at $x = 0$ is $f'(0) = \ln 3$.

35. (a) The population of Boomtown on January 1, 2000, was $P(40) \approx 22{,}285$.

    (b) $P'(40) = 50000 \cdot \ln(0.98) \cdot (0.98)^{40} \approx -450$. Thus, on January 1, 2000, the population was of Boomtown was *decreasing* at a rate of about 450 people per year.

37. (a) The average velocity of the particle over the interval $1 \le t \le e$ is
    $(x(e) - x(1))/(e - 1) = (1 - 0)/(e - 1) = 1/(e - 1)$ meters per second.

    (b) The instantaneous velocity of the particle at time t is $x'(t) = 1/t$. Thus, at time $t = e$, the instantaneous velocity of the particle is $x'(e) = 1/e$ meters per second.

39. Note that $f'(x) = \dfrac{1}{2\sqrt{x}} - \dfrac{1}{x} = \dfrac{\sqrt{x} - 2}{2x}$. Therefore, f is decreasing over the interval $(0, 4)$ and increasing over the interval $(4, \infty)$. This implies that f achieves its global minimum vlaue at $x = 4$.

41. Note that $f'(x) = \dfrac{1}{x} - \dfrac{e^x}{e} = \dfrac{e - xe^x}{e \cdot x}$. This implies that f is increasings on the interval $(0, 1)$ and decreasing on the interval $(1, \infty)$. Thus, f achieves its global maximum value at $x = 1$.

SECTION 2.6 DERIVATIVES OF EXPONENTIAL AND LOGARITHMIC FUNCTIONS; MODELING GROWTH

43. $y(x) = 100e^{0.1x}$

45. $y(x) = e^{(\ln 2)x} = 2^x$

47. The solution of the IVP $P' = 0.08P$, $P(0) = 2000$, is $P(t) = 2000e^{0.08t}$. Since $P(10) = 2000e^{0.8} \approx 4451.08$, a \$2000 deposit now will be worth approximately \$4451 after 10 years.

49. (a) $L(0) = a$ and $f(0) = 1$, so $a = 1$. $L'(0) = b$ and $f'(0) = 1$, so $b = 1$.

  (b) Since $Q(x) = L(x) + cx^2$, $Q(0) = L(0) = f(0)$ and $Q'(0) = L'(0) = f'(0)$. Now, $Q''(x) = 2c$ and $f''(x) = e^x$, so $Q''(0) = f''(0) \implies c = 1/2$.

  (c) Since $C(x) = Q(x) + dx^3$, $C(0) = Q(0) = f(0)$, $C'(0) = Q'(0) = f'(0)$, and $C''(0) = Q''(0) = f''(0)$. Now, $C'''(x) = 6d$ and $f'''(x) = e^x$, so $f'''(0) = 1 = C'''(0) = 6d \implies d = 1/6$.

  (e) $[-0.48318, 0.41622]$

  (f) $[-0.90509, 0.78672]$

  (g) $[-1.3225, 1.1683]$

51. Since $y' = (\ln 2)2^x$, the line tangent to the curve $y = 2^x$ at $x = 0$ has slope $\ln 2$ and passes through the point $(0, 1)$. Therefore, it is described by the equation $y = (\ln 2)x + 1$. Setting $y = 0$ and solving this equation for x, we find that the x-intercept is $x = -1/\ln 2$.

53. The slope of the line tangent to the curve $y = x - \ln x$ at $x = a$ is $y' = 1 - 1/a$. Since the tangent line passes through the point $(a, a - \ln a)$, it is described by the equation
$y = (1 - 1/a)(x - a) + (a - \ln a) = (1 - 1/a)x + (\ln a - 1)$. This line passes through the origin only if $\ln a = 1$. Thus, if $a = e$, the line tangent to the curve $y = x - \ln x$ at $x = a$ passes through the origin.

55. (a) If $P(t) = 1000 + Ce^{0.1t}$, then $P'(t) = 0.1Ce^{0.1t}$. Also,
$0.1P - 100 = (100 + 0.1Ce^{0.1t}) - 100 = 0.1Ce^{0.1t}$. Therefore, $P'(t) = 0.1P(t) - 100$.

  (b) The solution of the IVP $P' = 0.1P - 100$, $P(0) = 2000$, is $P(t) = 1000 + 1000e^{0.1t}$. Therefore, $P(10) = 1000 + 1000e^1 \approx 3718.28$. That is, the initial deposit of \$2000 will grow to approximately \$3718.

57. $y(x) \cdot y'(x) = y(x) \cdot (Ae^x - 1) = y(x) \cdot (y(x) + x) = (y(x))^2 + x \cdot y(x)$

59. $y'(x) = -A/x^2 + 1/x$ and $y''(x) = 2A/x^3 - 1/x^2$. Thus, $x^2 y''(x) + 2xy'(x) = 2A/x - 1 - 2A/x + 2 = 1$.

61. (a) The distance between the point $(0, 0)$ and a point (x, y) on the curve $y = e^{x/2}$ is
$D(x) = \sqrt{x^2 + (e^{x/2})^2} = \sqrt{x^2 + e^x}$. Since distance is a nonnegative quantity, minimizing the distance between two points is equivalent to minimizing the square of the distance between the two points. Thus, we are led to the problem of minimizing the function $f(x) = (D(x))^2 = x^2 + e^x$. Since $f'(x) = 2x + e^x = 0$ if $x \approx -0.35173$ and $f''(x) = 2 + e^x > 0$ for all $x \in \mathbb{R}$, the distance function is minimized if $x \approx -0.35173$.

  (b) The minimum distance is approximately 0.9095.

63. Let $B(t)$ be the weight of The Blob at time t hours after noon on Wednesday. Since The Blob grows at a rate proportional to its size, its growth is described by the differential equation $B' = kB$ and, therefore, $B(t) = Ae^{kt}$ for some numbers A and k. Since $B(0) = 1$ and $B(4) = 4$, $A = 1$ and $k = \frac{1}{4}\ln 4$. Thus, The Blob will weigh 3×10^{15} at time $t = 4\ln(3 \times 10^{15})/\ln 4 \approx 102.83$ hours (or, equivalently, in about 4.28 days).

65. (a) Since the bottle becomes full at $t = 24$ hours and the volume of the culture doubles every hour, the bottle must be half-full at time $t = 23$ hours.

(b) Let $V(t)$ be the volume of the culture at time t. Then, since the bottle is full at time $t = 24$, the bottle is less than 1% full if $V(t) < 0.01V(24)$ or $V(t)/V(24) < 0.01$.

The volume of the culture grows at a rate proportional to the amount present, so by Theorem ?? $V(t) = V(0)e^{kt}$. Since $V(1) = 2V(0)$, $k = \ln 2$ so $V(t) = V(0)2^t$. It follows that the bottle is less than 1% full when $2^{t-24} < 0.01$; that is, when $t < 24 + \log_2(0.01) \approx 17.356$ hours. Thus, the bottle is less than 1% full approximately 72% of the time.

67. Let $C(t)$ be the amount of C^{14} present in the skeletal fragments at time t (in years). For convenience, let t_{now} be the time the analysis was conducted, and $t = 0$ correspond to the date of death. At the time the analysis was performed, $C(t_{now})/C(0) = 0.0625$. Because the rate of decay of C^{14} is proportional to the amount of C^{14} present, $C'(t) = kC(t)$ for some constant k. The solution of this differential equation is $C(t) = C(0)e^{kt}$.

The value of k can be determined from the information given about the half-life of C^{14}. Since the half-life of C^{14} is 5728 years, $0.5 = e^{5728k}$ implies that $k = -(\ln 2)/5728$. Therefore, $C(t_{now})/C(0) = 0.0625 = e^{-(\ln 2/5728)t_{now}}$. From this it follows that $t_{now} = -\ln(0.0625) \cdot 5728/\ln 2 \approx 22{,}912$. Thus, the fragments date from about 22,912 years ago.

69. $\lim\limits_{t \to 0} \dfrac{e^t - 1}{t} = \dfrac{d}{dt}e^t \bigg|_{t=0} = 1$

§2.7 Derivatives of Trigonometric Functions: Modeling Oscillation

1. $f'(0) = \lim_{h \to 0} \dfrac{f(0+h) - f(0)}{h} = \lim_{h \to 0} \dfrac{\sin h - \sin 0}{h} = \lim_{h \to 0} \dfrac{\sin h}{h}$ (since $\sin 0 = 0$).

3. $\lim_{x \to 0} \dfrac{\sin x}{x} = 1$

5. (a) The graph of g is the graph of f shifted left by $\pi/2$ units. Thus, $g'(x) = f'(x + \pi/2) = \cos(x + \pi/2)$.

   (b) Applying the trigonometric identity $\cos(x + \pi/2) = -\sin x$ to the result in part (a) leads to $g'(x) = -\sin x$.

   (c) The trigonometric identity $\sin(x + \pi/2) = \cos x$ implies that $g(x) = h(x)$. Therefore, $h'(x) = g'(x) = -\sin x$.

7. The function $G(x) = -4\cos x + 3\sin x + C$, where C is a constant, is an antiderivative of g.

9. Any function of the form $x^3/3 + \cos x + C$, where C is a constant, is an antiderivative of $x^2 - \sin x$. The condition $F(0) = 1$ implies that $C = 0$, so $F(x) = x^3/3 + \cos x$.

11. $g''(x) = -\sin x$. Thus, g has inflection points at $x = -\pi$, $x = 0$, and $x = \pi$.

13. Since $g'(x) = 1 + \cos x \geq 0$ for all x, g is increasing throughout the interval I. Thus, g has no local minima in I. (g achieves its minimum value at $x = -5$.)

15. $y' = \sqrt{k}A\cos(\sqrt{k}x) - \sqrt{k}B\sin(\sqrt{k}x)$ so
    $y'' = -kA\sin(\sqrt{k}x) - kB\cos(\sqrt{k}x) = -k\big(A\sin(\sqrt{k}x) + B\cos(\sqrt{k}x)\big) = -ky$.

17. $f'(x) = 2\cos x$

19. $f'(x) = 3\cos(3x)$

21. $f'(x) = 3\cos x + 2\sin x$

23. $f''(x) = -2\sin x$

25. $f''(x) = -9\sin(3x)$

27. $f''(x) = -3\sin x + 2\cos x$

29. $F(x) = -2\cos x + C$

31. $F(x) = -\frac{1}{3}\cos(3x) + C$

33. $F(x) = -3\cos x - 2\sin x + C$

35. (a) $f'(x) = 1 + \sin x$. Thus, f has a stationary point (i.e., $f'(x) = 0$) at $x = 3\pi/2$.

    (b) $f'(x) = 1 + \sin x$. Thus, f is increasing ($f'(x) \geq 0$) everywhere in the interval $[0, 2\pi]$.

    (c) Since f is increasing everywhere in the interval $[0, \pi]$, the maximum value of f over the interval is $f(\pi) = \pi + 1$ and the minimum value of f over the interval is $f(0) = -1$.

    (d) f is concave down where $f''(x) < 0$. Since $f''(x) = \cos x$, so f is concave down on the interval $(\pi/2, 3\pi/2)$.

    (e) The function f is increasing most rapidly at the point(s) where $f'(x) = 1 + \sin x$ achieves its maximum value. This occurs at $x = \pi/2$; $f'(\pi/2) = 2$.

37. $f'(x) = 1/2 + \cos x$
$f''(x) = -\sin x$
local maximum at $(2\pi/3, \sqrt{3}/2 + \pi/3)$
local minimum at $(4\pi/3, -\sqrt{3}/2 + 2\pi/3)$
global maximum at $(2\pi, \pi)$
global minimum at $(0, 0)$
inflection point at $(\pi, \pi/2)$

39. $f'(x) = \frac{3}{2} + \cos x$
$f''(x) = -\sin x$
no local extrema
global maximum at $(2\pi, 3\pi)$
global minimum at $(0, 0)$
inflection point at $(\pi, 3\pi/2)$

41. $f'(x) = 2\cos x - \sqrt{3}$
$f''(x) = -2\sin x$
local (and global) maximum at $(\pi/6, 1 - \pi\sqrt{3}/6)$
local (and global) minimum at $(11\pi/6, -1 - 11\pi\sqrt{3}/6)$
inflection point at $(\pi, -\pi\sqrt{3})$

43. Note that $f'(x) = \cos x + \cos(3x)$.

   (a) $f'(x) = 0$ at $x = \pi/4$, $x = \pi/2$, and $x = 3\pi/4$.

   (b) $f'(x) < 0$ on the intervals $(\pi/4, \pi/2)$ and $(3\pi/4, \pi)$.

   (c) The maximum value of f over the interval $[0, \pi]$ is $f(\pi/4) = 2\sqrt{2}/3$. The minimum value of f over the interval $[0, \pi]$ is $f(0) = 0$.

45. (a) $f'(x) = -\sin x - \sin(2x)$ changes sign at $x = 2\pi/3$, $x = \pi$, and $x = 4\pi/3$. f has a local maximum at $x = \pi$ and local minima at $x = 2\pi/3$ and $x = 4\pi/3$.

   (b) $f''(x) = -\cos x - 2\cos(2x)$. f has inflection points near $x = 0.93593$, $x = 2.57376$, and $x = 3.70942$.

47. Many answers are possible. At $x = -1$, for instance, the graph of $f(x) = \sin(x^2)$ is decreasing so $f'(-1) < 0$ but $\cos((-1)^2) = \cos 1 > 0$.

49. (a) $f(0) = \sin 0 = 0 = L(0) = a$ and $f'(0) = \cos 0 = 1 = L'(0) = b$. Therefore, $a = 0$ and $b = 1$.

   (b) $Q(0) = L(0) = f(0)$ and $Q'(0) = L'(0) = f'(0)$. Now, $f''(0) = -\sin 0 = 0$ and $Q''(0) = 2c$, so $c = 0$.

   (c) $C(0) = Q(0) = f(0)$, $C'(0) = Q'(0) = f'(0)$, and $C''(0) = Q''(0) = f''(0)$. Now, $f'''(0) = -\cos 0 = -1$ and $C'''(0) = 6d$, so $d = -1/6$.

   (e) $[-0.85375, 0.85375]$

   (f) $[-0.85375, 0.85375]$

   (g) $[-1.6654, 1.6654]$

51. $\cos(x+k) = \dfrac{d}{dx}\sin(x+k) = \dfrac{d}{dx}(\sin x \cos k + \cos x \sin k) = \cos x \cos k - \sin x \sin k$

53. The general solution of the DE is $f(x) = \frac{1}{2}x^4 - \frac{1}{4}\cos(4x) + C$, where C is a constant. The initial condition $f(0) = 1$ implies that $C = \frac{5}{4}$. Thus, the solution of the IVP is $f(x) = \dfrac{1}{2}x^4 - \dfrac{1}{4}\cos(4x) + \dfrac{5}{4}$.

SECTION 2.7 DERIVATIVES OF TRIGONOMETRIC FUNCTIONS: MODELING OSCILLATION

55. If $h(x) = A\cos(\sqrt{k}x) + B\sin(\sqrt{k}x)$, then $h'(x) = -\sqrt{k}A\sin(\sqrt{k}x) + \sqrt{k}B\cos(\sqrt{k}x)$ and $h''(x) = -kA\cos(\sqrt{k}x) - kB\sin(\sqrt{k}x) = -kh(x)$. Thus, $h''(x) = -4h(x) \implies k = 4$, $h(0) = 2 \implies A = 2$, and $h'(0) = 8 \implies B = 4$.

57. First, note that $f''(x) = -f(x)$. Therefore, $f^{(4)}(x) = \left(f''(x)\right)'' = \left(-f(x)\right)'' = -f''(x) = f(x)$

59. (a) Let $h(x) = f(x) - g(x)$. Since $h'(x) = -\sin x - \sqrt{3}\cos x$, h has only one stationary point in the interval $[0, \pi]$ — at $x = 2\pi/3$. Now, $|h(0)| = 1$, $|h(2\pi/3)| = 2$, and $|h(\pi)| = 1$ so that maximum vertical distance between the curves is 2.

    (b) $f'(2\pi/3) = g'(2\pi/3) = -\sqrt{3}/2$

61. Assume that $0 < x < \pi/2$. Then, the coordinates of the rectangle are $(\pi/2 - x, 0)$, $(\pi/2 + x, 0)$, $(\pi/2 - x, 2\sqrt{2}\sin(\pi/2 - x)) = (\pi/2 - x, 2\sqrt{2}\cos x)$, and $(\pi/2 + x, 2\sqrt{2}\sin(\pi/2 + x)) = (\pi/2 + x, 2\sqrt{2}\cos x)$. Therefore, the perimeter of the rectangle is $P(x) = 4x + 4\sqrt{2}\cos x$. On the interval $(0, \pi/2)$, P achieves its maximum value at $x = \pi/4$; $P(\pi/4) = \pi + 4$.

63. (a) Since $f'(x) = \cos x$, the line tangent to the curve $y = f(x)$ at $x = 0$ has slope 1 and passes through the point $(0, 0)$. Thus, an equation of the tangent line is $y = x$.

    (b) Since $f'(x) = \cos x$, the line tangent to the curve $y = f(x)$ at $x = \pi/3$ has slope $1/2$ and passes through the point $(\pi/3, \sqrt{3}/2)$. Thus, an equation of the tangent line is $y = (x - \pi/3)/2 + \sqrt{3}/2$

    (c) The tangent line in part (a) yields the better approximation. Since the magnitude of f'' is less on the interval $(0, \pi/6)$ than on the interval $(\pi/6, \pi/3)$, this result could have been predicted. (The slope of the f-graph is changing less rapidly on the first interval than on the second.)

65. $\lim\limits_{x \to \pi} \dfrac{\sin x}{x - \pi} = \dfrac{d}{dx}\sin x \Big|_{x=\pi} = \cos \pi = -1$

67. $\lim\limits_{h \to 0} \dfrac{2\cos(5\pi/4 + h) + \sqrt{2}}{h} = \dfrac{d}{dx} 2\cos x \Big|_{x=5\pi/4} = -2\sin(5\pi/4) = \sqrt{2}$

§2.8 Chapter summary

1. No — $\lim_{x \to -1} f(x)$ does not exist because $\lim_{x \to -1^+} f(x) \neq \lim_{x \to -1^-} f(x)$.

3. Since $\frac{d}{dx}x^2 = 2x$, $\frac{d}{dx}e^x = e^x$, and $\frac{d}{dx}4 = 0$, $f'(x) = 2x - 3e^x$.

5. $f''(x) = 2 - 3e^x$, so $f''(2) = 2 - 3e^2 \approx -20.2 < 0$. This implies that f is concave down at $x = 2$.

7. Since $h'(w) = 0.5 - \cos w$, h has the form $h(w) = 0.5w - \sin w + C$, where C is a constant. Since $h(0) = 1$, $C = 1$. Thus, $h(w) = 0.5w - \sin w + 1$.

9. Yes. $h''(w) = \sin w \implies h''(\pi/2) = 1 > 0$. Thus, h is concave up at $w = \pi/2$.

11. $\lim_{x \to 1} \sqrt{x-1}$ does not exist because $\sqrt{x-1}$ is not defined when $x < 1$. (It is true, however, that $\lim_{x \to 1^+} \sqrt{x-1} = 0$.)

13. Yes. Let F denote the antiderivative of f. Then, $F''(x) = f'(x) = -2x^{-3} - \frac{2}{3}x^{-1/3}$ so $F''(1) = -8/3 < 0$. This implies that F is concave down at $x = 1$.

15. The limit expression is the definition of $g'(4)$. Thus, $\lim_{x \to 4} \frac{g(x) - g(4)}{x - 4} = g'(4) = 3 \cdot 4 + 4/4 - 5 = 8$.

17. Yes, because $g''(x) = 3 - 4/x^2$ changes sign (from negative to positive) at $x = 2\sqrt{3}/3$.

19. Since $g(1) = 2$, $g(x) = 3x^2/2 + 4\ln x - 5x + 11/2$. Therefore, the average rate of change of g over the interval $[6, 9]$ is

$$\frac{g(9) - g(6)}{9 - 6} = \frac{82 + 4\ln 9 - 59/2 - 4\ln 6}{3} = \frac{105}{6} + \frac{4}{3}(\ln 3 - \ln 2).$$

21. Let $C(t)$ be the concentration (in ppm) at time t (in years). Then,

$$C'(1995) \approx \frac{C(1996) - C(1995)}{1996 - 1995} = 362.69 - 360.91 = 1.78 \text{ ppm per year.}$$

[NOTE: Other answers are possible.]

23. $f(x) = \sqrt{x^3} + 4 = x^{3/2} + 4 \implies f'(x) = \frac{3}{2}x^{1/2} = 3\sqrt{x}/2$

25. $f(x) = 2\ln x + 3\cos x \implies f'(x) = 2/x - 3\sin x$

27. $f(x) = \pi - x^{-3/2} \implies F(x) = \pi x - \frac{x^{-1/2}}{-1/2} = \pi x + 2/\sqrt{x} + C$

29. $h(z) = 6z^5 + \frac{5}{z^{10}} = 6z^5 + 5z^{-10} \implies H(z) = 6\frac{z^6}{6} + 5\frac{z^{-9}}{-9} + C = z^6 - \frac{5}{9z^9} + C$

31. No; g is concave down at $x = 2$ because $g''(2) = -7 < 0$.

33. $\lim_{x \to \sqrt{2}} \frac{x^4 - 2}{x - \sqrt{2}} = f'(\sqrt{2})$, where $f(x) = x^4$. Since $f'(x) = 4x^3$, $\lim_{x \to \sqrt{2}} \frac{x^4 - 2}{x - \sqrt{2}} = 4 \cdot (\sqrt{2})^3 = 4 \cdot 2^{3/2}$.

35. A function F attains its maximum value at a stationary point or at an endpoint of the interval. Now, $F' = f$ so F has only one stationary point in the interval $[0, 10]$, at 1. Finally, since f is positive on the interval $[0, 1)$ and negative on the interval $(1, 10]$, $F(x)$ attains its maximum value at $x = 1$; $F(x)$ attains its minimum value at either $x = 0$ or $x = 10$.

37. Yes, $F(x)$ is concave up at $x = 3$ because $f(x) = F'(x)$ is increasing at $x = 3$.

SECTION 2.8 CHAPTER SUMMARY

39. If F is an antiderivative of f, $F' = f$. Therefore, $F'(x) = f(x) = x \sin(e^x)$.

41. $g(w) = \pi w^2 + 4\sqrt[3]{w^2} - 1/w = \pi w^2 + 4w^{2/3} - w^{-1} \implies g'(w) = 2\pi w + \frac{8}{3}w^{-1/3} + 1/w^2$

43. $P(t) = 3t^2 + 4\sqrt{t} + 5 = 3t^2 + 4t^{1/2} + 5 \implies P'(t) = 6t + 2t^{-1/2} = 6t + 2/\sqrt{t}$

45. $\dfrac{dh}{dw} = h'(w) = \cos w + \sin w$

47. No. $y'(x) = 10x^4 \neq 10y/x = 10 \cdot (2x^5)/x = 20x^4$.

49. No, because $f'(3) = \lim\limits_{h \to 0} \dfrac{f(3+h) - f(3)}{h} = -2$.

51. (a) The graph of g is the graph of f shifted to the right by 2 units. Therefore, the graph of g' is the graph of f' shifted to the right by 2 units. Since $f'(x) = 4x^3$, this implies that $g'(x) = 4(x-2)^3$.
    (b) Using reasoning similar to that in part (a), $h'(x) = k(x-c)^{k-1}$.
    (c) $k(x) = (4-x)^3 = -(x-4)^3$. Therefore, $k'(x) = -3(x-4)^2 = -3(4-x)^2$.

53. $F(\theta) = \sin\theta \implies F'(\theta) = \cos\theta \implies F''(\theta) = -\sin\theta$

55. $U(r) = \ln r \implies U'(r) = 1/r = r^{-1} \implies U''(r) = -r^{-2} = -1/r^2$

57. $\lim\limits_{x \to 1} \dfrac{x^{1000} - 1}{x - 1} = f'(1)$ where $f(x) = x^{1000}$. Since $f'(x) = 1000x^{999}$, $\lim\limits_{x \to 1} \dfrac{x^{1000} - 1}{x - 1} = 1000$.

59. $f'(x) = 12x^3 - 24x^2 = 12x^2(x-2)$. Since $f'(x) \leq 0$ if $x < 2$ and $f'(x) > 0$ if $x > 2$, f is decreasing on the interval $(-\infty, 2)$ and increasing on the interval $(2, \infty)$. This implies that $f(x) \geq f(2) = 1 > 0$ for every x.

61. The weight's velocity and acceleration at time t are $y'(t) = \frac{\pi}{2}\sin(\pi t)$ and $y''(t) = \frac{\pi^2}{2}\cos(\pi t)$, respectively. Thus, At time $t = 1/3$, the weight's velocity and acceleration are $y'(1/3) = \sqrt{3}\pi/4$ and $y''(1/3) = \pi^2/4$, respectively.

63. $g'(x) = f'(x + \pi)$ so $g'(0) = f'(\pi) = \pi^2 \cos\pi = -\pi^2$

65. $j'(x) = 2f'(x) = 2x^2 \cos x$

67. The function G is an antiderivative of g if $G' = g$. Thus, $G(x) = F(x) + 3x$ is an antiderivative of g.

69. The function J is an antiderivative of j if $J' = j$. Thus, $J(x) = 3F(x)$ is an antiderivative of j.

§3.1 Algebraic Combinations: The Product and Quotient Rules

1. (a) Observe that $p(x) = f(x) \cdot g(x) = (1-x)(1+x) = 1 - x^2$, $f'(x) = -1$, and $g'(x) = 1$. It is clear from a graph of p that $p'(x) > 0$ if $x < 0$, but $f'(x) \cdot g'(x) = -1$ for any x. Thus, $p'(x) \neq f'(x) \cdot g'(x)$.

   (b) $f'(2)/g'(2) = -1$ but $q'(2) \approx -0.222$.

3. $g'(t) = \dfrac{(-\sin t)t^4 - 4t^3 \cos t}{t^8} = -\dfrac{t \sin t + 4 \cos t}{t^5}$

5. $r'(x) = \left(\dfrac{1}{2^x}\right)' = \dfrac{0 \cdot 2^x - 1 \cdot 2^x \ln 2}{(2^x)^2} = -\dfrac{\ln 2}{2^x} = -2^{-x} \ln 2.$

7. $g'(z) = e^z \tan z + e^z \sec^2 z$

9. (i) $h'(x) = 1 \cdot x^{-1} + x \cdot (-x^{-2}) = x^{-1} - x^{-1} = 0$
   (ii) $h(x) = x \cdot x^{-1} = 1 \implies h'(x) = 0$

11. (i) $h'(x) = \frac{1}{2}x^{-1/2} \cdot x^{1/3} + x^{1/2} \cdot \frac{1}{3}x^{-2/3} = \left(\frac{1}{2} + \frac{1}{3}\right)x^{-1/6} = \frac{5}{6}x^{-1/6}$
    (ii) $h(x) = x^{1/2} \cdot x^{1/3} = x^{5/6} \implies h'(x) = \frac{5}{6}x^{-1/6}$

13. (i) $h'(x) = e^x \cdot e^{2x} + e^x \cdot 2e^{2x} = (1+2)e^{3x} = 3e^{3x}$
    (ii) $h(x) = e^x \cdot e^{2x} = e^{3x} \implies h'(x) = 3e^{3x}$

15. (i) $h'(x) = \dfrac{2x \cdot x^3 - x^2 \cdot 3x^2}{x^6} = -x^4/x^6 = -1/x^2 = -x^{-2}$
    (ii) $h(x) = \dfrac{x^2}{x^3} = x^{-1} \implies h'(x) = -x^{-2}$

17. (i) $h'(x) = \dfrac{0 \cdot e^{2x} - 2 \cdot e^{2x}}{(e^{2x})^2} = -2e^{2x}/e^{4x} = -2e^{-2x}$
    (ii) $h(x) = \dfrac{1}{e^{2x}} = e^{-2x} \implies h'(x) = -2e^{-2x}$

19. (i) $h'(x) = \dfrac{-3x^2(1-x) - (1-x^3)(-1)}{(1-x)^2} = \dfrac{-3x^2 + 2x^3 - 1}{(1-x)^2} = 1 + 2x$
    (ii) $h(x) = \dfrac{1-x^3}{1-x} = 1 + x + x^2 \implies h'(x) = 1 + 2x$

21. $f'(x) = 1 \cdot \sin x + x \cos x = \sin x + x \cos x$

23. $f'(x) = \frac{1}{2}x^{-1/2} \cdot \ln x + x^{1/2} \cdot x^{-1} = \left(\frac{1}{2} \ln x + 1\right)/\sqrt{x}$

25. $f'(x) = 2e^{2x} \cdot \ln x + e^{2x} \cdot x^{-1} = (2 \ln x + x^{-1})e^{2x}$

27. $f'(x) = 2x \cdot e^x + (x^2 + 3) \cdot e^x = (x^2 + 2x + 3)e^x$

29. $f'(x) = \dfrac{(\cos x) \cdot x^2 - (\sin x) \cdot (2x)}{(x^2)^2} = \dfrac{x \cos x - 2 \sin x}{x^3}$

31. $f'(x) = \dfrac{2(x+1) \cdot (x^2 + 1) - (x+1)^2 \cdot (2x)}{(x^2+1)^2} = \dfrac{2(1-x^2)}{(x^2+1)^2}$

33. $f'(x) = -3 \csc^2(3x)$

35. $f'(x) = -6 \csc(2x) \cot(2x)$

37. Yes, $G'(x) = \sec^2 x$.

39. Yes, $G'(x) = e^{2x} \sin x - 2e^{2x} \cos x$.

41. Yes, $G'(x) = 9x^2 \cdot \ln x + 3x^3 \cdot x^{-1} - 3x^2 = 9x^2 \ln x$.

43. $h'(x) = 2xf(x) + x^2 f'(x)$, $f(2) = -4$, $f'(2) = 0$, $f(4) = 0$, and $f'(4) = 4$. Thus, $h'(2) = 4f(2) + 4f'(2) = 4f(2) = -16$ and $h'(4) = 8f(4) + 16f'(4) = 64$.

45. $h''(x) = 2f(x) + 4xf'(x) + x^2 f''(x)$ so Since $f(-0.5) > 0$, $f'(-0.5) < 0$, $f''(-0.25) > 0$, $h''(-0.5) = 2f(-0.5) - 2f'(-0.5) + 0.25f''(-0.25) > 0$.

47. $m'(x) = \dfrac{(x^2+1)f'(x) - 2xf(x)}{(x^2+1)^2}$. Thus, $m'(0) = f'(0) = -4$ and $m'(2) = 16/25 = 0.64$.

49. $g'(x) = 3x^2 f(x) + x^3 f'(x) \implies g'(1) = 3f(1) + f'(1) = 3 \cdot 2 + 4 = 10$

51. $g''(x) = 6xf(x) + 6x^2 f'(x) + x^3 f''(x) \implies g''(1) = 6f(1) + 6f'(1) + f''(1) = 6 \cdot 2 + 6 \cdot 4 + (-2) = 34$

53. $g'(x) = \cos x f(x) + \sin x f'(x) \implies g'(0) = f(0) = 3$

55. Yes. $g'(x) = \cos x f(x) + \sin x f'(x) \implies g'(\pi/2) = f'(\pi/2) > 3$. Since $g'(\pi/2) > 0$, g is increasing at $x = \pi/2$.

57. No. $g''(x) = -\sin(x)f(x) + 2\cos(x)f'(x) + \sin(x)f''(x) \implies g''(\pi/2) = -f(\pi/2) + f''(\pi/2)$. Since $f'(x) > 0$ on the interval $[0, 2]$, $f(\pi/2) > f(0) = 3$. Since the graph of f' is decreasing at $x = \pi/2$, $f''(\pi/2) < 0$. Therefore, $g''(\pi/2) < 0$. This implies that g is concave down at $x = \pi/2$.

59. Yes. $h'(x) = (f(x) + f'(x))e^x \implies h'(1) = (f(1) + f'(1))e^1$. Since $f'(x) > 0$ on the interval $[0, 1]$, $f(1) > f(0) = 3$. Therefore, $h'(1) > 0$ which implies that h is increasing at $x = 1$.

61. $h''(x) = e^x (f(x) + 2f'(x) + f''(x))$. Since $f(0) = 3$, $f'(0) = 1$, and $f''(0) \approx 2.5$, $h''(0) \approx 7.5$.

63. Yes. $h''(x) = e^x (f(x) + 2f'(x) + f''(x))$. Now, $f''(2) \approx -1.5$, $f'(2) = 3$, and, since $f'(x) > 1$ on the interval $[0, 2]$, $f(2) > f(0) + 2 \cdot 1 = 5$, $g''(2) > 0$. This implies that g is concave up at $x = 2$.

65. $k'(x) = 2f(x) \cdot f'(x) \implies k'(1) = 2f(1) \cdot f'(1)$. Now, $f'(x) > 0$ on the interval $[0, 1]$ so $f(1) > f(0) = 3$ and $f'(1) = 3$, so $k'(1) = 6f(1) > 0$. This implies that k is increasing at $x = 1$.

67. $k''(x) = 2(f'(x))^2 + 2f(x) \cdot f''(x)$. Therefore, $k''(0) = 2(f'(0))^2 + 2f(0) \cdot f''(0) = 2 \cdot (1)^2 + 6f''(0) \approx 2 + 6 \cdot 2.5 = 17$.

69. (a) First, note that $f'(x) = (1+x)e^x$ and $L'(x) = a$. Since $f'(0) = 1$ and $f'(0) = L'(0)$, we conclude that $a = 1$. Since $f(0) = 0$ and $f(0) = L(0)$, we conclude that $b = 0$.

(b) First, note that $f''(x) = (2+x)e^x$, and $Q''(x) = 2c$. From part (a), $f(0) = Q(0) = L(0)$ and $f'(0) = Q'(0) = L'(0)$. Since $f''(0) = 2$, $f''(0) = Q''(0)$ if $c = 1$.

(d) $[-0.34375, 0.29334]$

(e) $[-0.62463, 0.54865]$

71. (a) $h'(x) = f'(x)g(x) + f(x)g'(x)$ so $h''(x) = f''(x)g(x) + 2f'(x)g'(x) + fg''(x)$.

(b) $h'''(x) = f'''(x)g(x) + 3f''(x)g'(x) + 3f'(x)g''(x) + f(x)g'''(x)$

73. $f(0) = 3 \implies A = 3$, so $f(x) = (3 + Bx)e^x$. Now, $f'(x) = (3 + B + Bx)e^x$ so $f'(x) - f(x) = Be^x$. Thus, f will be a solution of the IVP if $A = 3$ and $B = 2$.

75. $G'(x) = (Ax^2 + (2A+B)x + (B+C))e^x = g(x) \implies A = 1, B = -2,$ and $C = 2$.

77. (a) $f'(x) = (1 - \ln x)/x^2$. Since $f'(x) > 0$ on the interval $[1, e)$ and $f'(x) < 0$ on the interval (e, ∞), f achieves its maximum value at $x = e$. $f(e) = 1/e$.

(b) $f''(x) = \dfrac{-3 + 2\ln x}{x^3}$ so $f''(x)$ changes sign only at $x = e^{3/2}$.

79. (a) $f'(x) = e^x(\cos x - \sin x)$, so f has only one stationary point in the interval $[0, \pi]$, $x = \pi/4$. Since $f(\pi) < f(0) < f(\pi/4)$, over the interval $[0, \pi]$, f achieves its maximum value at $x = \pi/4$

(b) f achieves its minimum value minimum value over the interval $[0, \pi]$ at $x = \pi$.

(c) $f''(x) = -2e^x \sin x$, so f has inflection points at $x = 0$ and at $x = \pi$.

81. (a) $y'(x) = (1 - 2(x + C))e^{-2x}$. Therefore,

$$y' + 2y = (1 - 2(x + C))e^{-2x} + 2(x + C)e^{-2x} = e^{-2x}.$$

(b) From part (a), $y(x) = (x + C)e^{-2x}$ is a solution of the DE for any real number C. Since $y(0) = C$, it follows that the solution of the IVP is $y(x) = (x + 7)e^{-2x}$.

§3.2 Composition and the Chain Rule

1. (a) $h(x) = \cos(x^2) = (g \circ f)(x) = g(f(x))$
   (b) Since $f'(x) = 2x$ and $g'(x) = -\sin(x)$, $h'(x) = g'(f(x)) \cdot f'(x) = -2x\sin(x^2)$.

3. Let $h(x) = \cos(2x)$. Then, $h(x) = f(g(x))$, where $f(u) = \cos u$ and $g(w) = 2w$. Since $f'(u) = -\sin u$ and $g'(w) = 2$, the chain rule implies that $\dfrac{d}{dx}\cos(2x) = f'(g(x)) \cdot g'(x) = -2\sin(2x)$.

5. $f(x) = g(h(x))$, where $g(u) = u^{29}$ and $h(v) = v^2 + 3$. The chain rule implies that
   $f'(x) = g'(h(x)) \cdot h'(x) = 29(x^2 + 3)^{28} \cdot 2x = 58x(x^2 + 3)^{28}$

7. $f(x) = (g \circ h)(x)$, where $g(u) = \sqrt{u}$ and $h(v) = 2 + \sin x$. The chain rule implies that
   $f'(x) = g'(h(x)) \cdot h'(x) = \tfrac{1}{2}(2 + \sin x)^{-1/2} \cdot \cos x = \dfrac{\cos x}{2\sqrt{2 + \sin x}}$.

9. $f(x) = g(h(x))$, where $g(u) = \cos u$ and $h(v) = v^3$. Therefore, the chain rule implies that
   $f'(x) = g'(h(x)) \cdot h'(x) = -\sin(x^3) \cdot 3x^2 = -3x^2\sin(x^3)$.

11. $f(x) = (g \circ h)(x)$, where $g(u) = \ln u$ and $h(w) = 4w^2 + 3$. Therefore, the chain rule implies that
    $f'(x) = g'(h(x)) \cdot h'(x) = \dfrac{1}{4x^2 + 3} \cdot 8x = \dfrac{8x}{4x^2 + 3}$.

13. $f(x) = g(h(x))$, where $g(t) = \cos t$ and $h(w) = \sqrt{w} + e^{-w}$. Therefore, the chain rule implies that
    $f'(x) = g'(h(x)) \cdot h'(x) = -\sin\left(\sqrt{x} + e^{-x}\right) \cdot \left(\tfrac{1}{2}x^{-1/2} - e^{-x}\right) = -\left(\dfrac{1}{2}x^{-1/2} - e^{-x}\right)\sin\left(\sqrt{x} + e^{-x}\right)$.

15. $f(x) = (g \circ h \circ j)(x)$, where $g(t) = t^2$, $h(u) = \cos u$, and $j(w) = w^3$. Now, the chain rule implies that
    $f'(x) = (g \circ h)'(j(x)) \cdot j'(x) = g'((h \circ j)(x)) \cdot h'(j(x)) \cdot j'(x)$. Therefore,
    $f'(x) = 2\cos(x^3) \cdot (-\sin(x^3)) \cdot 3x^2 = -6x^2\cos(x^3)\sin(x^3)$.

17. $f(x) = g(h(x))$, where $g(v) = \sqrt{v}$ and $h(t) = 1 + 3t^2$. Therefore, the chain rule implies that
    $f'(x) = g'(h(x)) \cdot h'(x) = \tfrac{1}{2}(1 + 3x^2)^{-1/2} \cdot 6x = \dfrac{3x}{\sqrt{1 + 3x^2}}$.

19. $f(x) = g(h(x))$, where $g(w) = \ln w$ and $h(t) = t^2 + 3\sin t$. Therefore, the chain rule implies that
    $f'(x) = g'(h(x)) \cdot h'(x) = (x^2 + 3\sin x)^{-1} \cdot (2x + 3\cos x) = \dfrac{2x + 3\cos x}{x^2 + 3\sin x}$.

21. $h(1) = f(g(1)) = f(4) = 3$
    $h'(1) = f'(g(1))g'(1) = f'(4)g'(1) = 4 \cdot 3 = 12$

x	$f(x)$	$f'(x)$	$g(x)$	$g'(x)$	$h(x)$	$h'(x)$
1	1	2	4	3	3	12
2	2	1	3	4	4	12
3	4	3	1	2	1	4
4	3	4	2	1	2	1

23. $f(x)$ has a local maximum where $f'(x)$ changes sign from positive to negative. Since $f'(x) = (-\sin x)/(2 + \cos x)$ and $2 + \cos x > 0$ for all x, $f(x)$ has local maxima at $x = 2k\pi$, where $k = 0, \pm 1, \pm 2, \ldots$.

25. f has an inflection point at x if f'' changes sign at x. Since $f''(x) = -(1 + 2\cos x)/(2 + \cos x)^2$, f has an inflection point at those values of x for which $\cos x = -1/2$. Thus, f has infection points at $x = \frac{2\pi}{3} + 2k\pi$ and $x = \frac{4\pi}{3} + 2k\pi$ where $k = 0, \pm 1, \pm 2, \ldots$.

27. Yes. $g'(x) = 2xf'(x^2)$. Since, $g'(-1) = -2f'(1) > 0$, $g(x)$ is increasing at $x = -1$. [NOTE: $f'(1) < 0$ since f is decreasing at $x = 1$.]

29. (a) $h(-2) = f(g(-2)) = f(3) = 1/2$
    $h(1) = f(g(1)) = f(0) = 0$
    $h(3) = f(g(3)) = f(-1) = 1/4$

    (b) $h'(-2) = f'(g(-2))g'(-2) = f'(3) \cdot g'(-2) \approx -1/2$
    $h'(1) = f'(g(1)) \cdot g'(1) = f'(0) \cdot g'(1) \approx 0$
    $h'(3) = f'(g(3)) \cdot g'(3) = f'(-1) \cdot g'(3) \approx -1/8$

31. Since $h'(x) = f'(g(x))g'(x)$, h has stationary points where either $f'(g(x)) = 0$ or $g'(x) = 0$. Since $f'(0) = 0$ and $g(-4) = g(1) = g(5) = 0$, h has stationary points at $x = -4$, $x = 1$, and $x = 5$. Since $f'(2) = 0$ and $g(-3) = g(0) = 2$, h has stationary points at $x = -3$ and $x = 0$. Since $g'(-1.5) = g'(2.5) = 0$, h has stationary points at $x = -1.5$ and $x = 2.5$. Thus, h has stationary points at $x = -4, x = -3, x = -1.5, x = 0, x = 1, x = 2.5$, and $x = 5$.

33. Yes. $k'(-4) = g'(f(-4))f'(-4) = g'(4)f'(-4)$. Since $g'(4) > 0$ and $f'(-4) < 0$, $k'(-4) < 0$ which implies that k is decreasing at $x = -4$.

35. $k'(x) = -e^{-x}f'(e^{-x})$. Now, $e^{-(-1)} = e \approx 2.72$, $e^{-0} = 1$, and $e^{-1} \approx 0.37$, so $f'(e^1) < 0$, $f'(1) > 0$, and $f'(e^{-1}) > 0$. Therefore, $k'(-1) > 0$, $k'(0) < 0$ and $k'(1) < 0$. Thus, k is increasing at $x = -1$ and decreasing at $x = 0$ and $x = 1$.

37. No, because $F'(x) = (4x + 4x^3)e^{x^2} \neq f(x)$.

39. Yes, because $F'(x) = \frac{1}{2}(1 + x^2)^{-1/2} \cdot 2x = \frac{x}{\sqrt{1+x^2}}$.

41. $g'(x) = f'(x + \pi) = (x + \pi)\cos(x + \pi)$ so $g'(0) = \pi \cos \pi = -\pi$.

43. $j'(x) = 2xf'(x^2) + 3 = 2x \cdot x^2 \cos(x^2) + 3 = 2x^3 \cos(x^2) + 3$

45. (a) The tangent line passes through the point $(\pi/2, 17)$ and has slope $f'(\pi/2) = 2$. Therefore, it is described by the equation $y = 2(x - \pi/2) + 17$.

    (b) No. Since $\pi/2 < 2$ and $f''(x) = 2(\cos x - x \sin x)e^{x \cos x} < 0$ if $\pi/2 \leq x \leq 2$ (i.e., the graph of f is concave down), $L(2) > f(2)$.

47. No. $g'(x) = f'(f(x)) \cdot f'(x) > 0$ for all x, so g is increasing for all x.

SECTION 3.3 IMPLICIT FUNCTIONS AND IMPLICIT DIFFERENTIATION

§3.3 Implicit Functions and Implicit Differentiation

1. Using implicit differentiation, the equation $5x^2 - 6xy + 5y^2 = 16$ implies $10x - 6y - 6xy' + 10yy' = 0$. Solving the latter equation for y' leads to the result $y' = \frac{6y - 10x}{10y - 6x}$.

   (a) The tangent line is horizontal where $y' = 0$ (i.e., where $6y = 10x$). Substituting $y = 5x/3$ into the equation for the ellipse, then solving, leads to the conclusion that the line $6y = 10x$ intersects the ellipse at the points $(3/\sqrt{5}, \sqrt{5})$ and $(-3/\sqrt{5}, -\sqrt{5})$. Therefore, these are points on the ellipse at which the the tangent line is horizontal.

   (b) The tangent line is vertical where $10y = 6x$ (or $y = 3x/5$). Since the line $10y = 6x$ intersects the ellipse at the points $(\sqrt{5}, 3/\sqrt{5})$ and $(-\sqrt{5}, -3/\sqrt{5})$, these are points on the ellipse at which the tangent line is vertical.

3. (a) $2x + 3y = 6 \implies y = 2 - 2x/3 \implies dy/dx = -2/3$.
   (b) $2x + 3y = 6 \implies 2 + 3y' = 0 \implies y' = dy/dx = -2/3$.

5. (a) $x^2 + y^2 = 1 \implies 2x + 2yy' = 0 \implies y' = -x/y$.
   (b) $y = \sqrt{1-x^2} \implies y' = -x/\sqrt{1-x^2} = -x/y$. Similarly, $y = -\sqrt{1-x^2} \implies y' = x/\sqrt{1-x^2} = -x/y$.

7. (a) The equation $y^2 - 2x^2 = 1 \implies y^2 = 1 + 2x^2$ defines the functions $f_1(x) = \sqrt{1+2x^2}$ and $f_2(x) = -\sqrt{1+2x^2}$.
   (c) $y^2 - 2x^2 = 1 \implies 2yy' - 4x = 0 \implies y' = 2x/y$.
   (d) The tangent line is horizontal if $y'(x) = 2x/y = 0$. Thus the tangent line is horizontal only at the points $(0, 1)$ and $(0, -1)$.
   (e) Substituting $x = c$ into the equation for the hyperbola, we find that the line $x = c$ intersects the hyperbola at the points $(c, \sqrt{1+2c^2})$, $(c, -\sqrt{1+2c^2})$. At these points, the hyperbola has slope $2c/\sqrt{1+2c^2}$ and $-2c/\sqrt{1+2c^2}$, respectively.

9. Using implicit differentiation, the equation $x^2 + xy + y^2 = 1 \implies 2x + y + xy' + 2yy' = 0$ so $y' = \frac{dy}{dx} = -\frac{2x+y}{x+2y}$. Similarly, the equation $x^2 - xy + y^2 = 1 \implies 2x - y - xy' + 2yy' = 0$ so $y' = \frac{dy}{dx} = \frac{-2x+y}{-x+2y}$.

11. On the curve $x^2 + xy + y^2 = 1$, the tangent lines at the points $(0, 1)$ and $(0, -1)$ have slope $-1/2$; the tangent lines at the points $(1, 0)$ and $(-1, 0)$ have slope -2. On the curve $x^2 - xy + y^2 = 1$, the tangent lines at the points $(0, 1)$ and $(0, -1)$ have slope $1/2$; the tangent lines at the points $(1, 0)$ and $(-1, 0)$ have slope 2.

13. For the curve $x^2 + xy + y^2 = 1$, $y' = -(2x+y)/(x+2y)$. This implies that this curve is vertical (i.e., y' is undefined) at the points on the curve where $y = -x/2$. Therefore, this curve is vertical at the points $(2/\sqrt{3}, -1/\sqrt{3})$ and $(-2/\sqrt{3}, 1/\sqrt{3})$. For the curve $x^2 - xy + y^2 = 1$, $y' = (-2x+y)/(-x+2y)$ so this curve is vertical at the points $(2/\sqrt{3}, 1/\sqrt{3})$ and $(-2/\sqrt{3}, -1/\sqrt{3})$.

15. (b) $2x + 4yy' = 0 \implies y' = -x/2y$

17. (b) $2yy' - 2x = 2xy^2 + 2x^2yy' \implies y' = (x + xy^2)/(y - x^2y)$.

19. $3x^2 + 3y^2y' = 4y + 4xy' \implies y' = (4y - 3x^2)/(3y^2 - 4x)$. Thus, $y'(2) = -1$ so the tangent line is described by the equation $y = -(x - 2) + 2 = 4 - x$.

21. $2(x^2 + y^2)^2 = 25xy \implies 4(x^2 + y^2)(2x + 2yy') = 25y + 25xy' \implies y' = (25y - 8x(x^2+y^2))/(8y(x^2+y^2) - 25x)$. Thus, the tangent line has slope $y'(1) = 2/11$ and is described by the equation $y = (2/11)(x-1) + 2 = (2x+20)/11$.

23. By implicit differentiation, $2yy' = 3x^2$ so $y' = 3x^2/(2y)$. Therefore, the line tangent to the curve at $(1, 1)$ is described by the equation $y = \frac{3}{2}(x - 1) + 1$. Similarly, the line tangent to the curve at $(1, -1)$ is described by the equation $y = \frac{-3}{2}(x - 1) - 1$.

25. Using implicit differentiation, $y'(x) = \dfrac{1 - y\sin(xy)}{1 + x\sin(xy)}$. This implies that the line tangent to the curve near the point $(0, 1)$ has slope $y'(0) = 1$. Therefore, using the tangent line approximation, $y(0.025) \approx y(0) + y'(0) \cdot 0.025 = 1.025$.

27. Using implicit differentiation, $1 = y'e^y \implies y' = e^{-y} = 1/e^y = 1/x$.

29. Using implicit differentiation, $y^2 = 1 + Ce^{-2x} \implies 2yy' = -2Ce^{-2x}$. Now, substituting into the DE,

$$y(y' + y) = yy' + y^2 = -Ce^{-2x} + (1 + Ce^{-2x}) = 1.$$

§3.4 Inverse Functions and their Derivatives; Inverse Trigonometric Functions

1. $\arcsin(1) = \pi/2$ since $\sin(\pi/2) = 1$

3. $\arccos(-1) = \pi$ since $\cos(\pi) = -1$

5. $\arctan(1) = \pi/4$ since $\tan(\pi/4) = 1$

7. Draw a right triangle with sides x and $\sqrt{1-x^2}$ such that the side of length $\sqrt{1-x^2}$ is opposite the angle θ. Then, $\cos\theta = x$ and $\sin\theta = \sqrt{1-x^2}$. Since $\theta = \arccos x$, $\sin(\arccos x) = \sqrt{1-x^2}$

9. Draw a right triangle with sides 1 and x such that the side of length x is opposite the angle θ. Then, $\tan\theta = x$ and $\cos\theta = 1/\sqrt{1+x^2}$. Since $\theta = \arctan x$, $\cos(\arctan x) = 1/\sqrt{1+x^2}$.

11. Draw a right triangle with sides 1 and $2x$ such that the side of length $2x$ is opposite the angle θ. Then, $\tan\theta = 2x$ and $\cos\theta = 1/\sqrt{1+4x^2}$. Since $\theta = \arctan(2x)$, $\cos(\arctan(2x)) = 1/\sqrt{1+4x^2}$.

13. Since $(\arctan u)' = (1+u^2)^{-1}$, the chain rule implies that $f'(x) = 2/(1+4x^2)$.

15. Using the chain rule, $f'(x) = \frac{1}{2}(\arcsin x)^{-1/2}(\arcsin x)' = \frac{1}{2}\left((1-x^2)\arcsin x\right)^{-1/2}$.

17. Using the chain rule, $f'(x) = e^x/\sqrt{1-e^{2x}}$.

19. Using the chain rule, $f'(x) = \left(x(1+(\ln x)^2)\right)^{-1}$

21. $f'(x) = \dfrac{1}{2|x/2|\sqrt{(x/2)^2-1}} = \dfrac{1}{|x|\sqrt{x^2/4-1}} = 2/\sqrt{x^4-4x^2}$.

23. $f'(x) = 2x\arctan(\sqrt{x}) + x^2\dfrac{1}{2\sqrt{x}(1+x)} = 2x\arctan(\sqrt{x}) + x^{3/2}/(2+2x)$

25. Since $\cos(\pi/3) = 1/2$ and the cosine function is 2π-periodic, $\cos x = 1/2$ if $x = \pi/3$ or $x = \pi/3 + 2\pi = 7\pi/3$. Since the cosine function is an even function, $\cos x = 1/2$ if $x = -\pi/3$ or $x = -7\pi/3$.

27. Since the point $(-3, 5)$ is on the graph of f, $f(-3) = 5$. This implies that $f^{-1}(5) = -3$ so the point $(5, -3)$ is on the graph of f^{-1}.

29. Reflecting a horizontal line about the line $y = x$ produces a vertical line, and vice versa.

31. If $f(g(x)) = x$, then $f'(g(x))g'(x) = 1$ so $g'(x) = 1/f'(g(x))$. Thus, since $f'(x) < 0$ for all x, $g'(x) < 0$ for all x.

33. The function f is one-to-one (or passes the horizontal line test), so it has an inverse. Since f is a linear function, its inverse can be found explicitly: $g(y) = (y-a)/b$.

35. If $y = f(u) = a + bu$, then, $u = (y-a)/b = g(y)$. Thus, $g(x) = (x-a)/b$.

37. The range of the arcsine function is $[-\pi/2, \pi/2]$, an interval that does not include 5.

39. Using the chain rule, $\left(\arcsin(\sin x)\right)' = \dfrac{\cos x}{\sqrt{1-\sin^2 x}} = \dfrac{\cos x}{|\cos x|}$.

    [NOTE: $\sin^2 x + \cos^2 x = 1 \implies \cos^2 x = 1 - \sin^2 x \implies |\cos x| = \sqrt{1-\sin^2 x}$.]

41. $(\arctan x)' = (1+x^2)^{-1}$, so the domain of the derivative of the arctangent function is $(-\infty, \infty)$. It is the same as the domain of the arctangent function.

43. (b) The lines $y = -\pi/2$ and $y = \pi/2$ are horizontal asymptotes of the arctangent function.

(c) f is an odd function.

(d) f is increasing on $(-\infty, \infty)$ because f' is positive on this interval.

(e) f is concave up on $(-\infty, 0)$ and concave down on $(0, \infty)$. f has an inflection point at $x = 0$.

45. (b) f is neither even nor odd. The graph of f is not symmetric about the y-axis and it is not symmetric about the origin.

(c) f is decreasing on $(-1, 1)$ because $f'(x) = \dfrac{-1}{\sqrt{1-x^2}}$ is negative on this interval.

(d) $f''(x) = \dfrac{-x}{(1-x^2)^{3/2}}$ so $f''(x) < 0$ on the interval $(-1, 0)$ and $f''(x) > 0$ on the interval $(0, 1)$. Thus, f is concave up on $(-1, 0)$ and concave down on $(0, 1)$. Since f'' changes sign at $x = 0$, f has an inflection point at $x = 0$.

47. (b) f is neither even nor odd. The graph of f is not symmetric about the y-axis and it is not symmetric about the origin.

(c) Since $f'(x) = \dfrac{1}{|x|\sqrt{x^2-1}} > 0$ if $-\infty < x < -1$ or $1 < x < \infty$), f is increasing on $(-\infty, -1) \cup (1, \infty)$.

(d) Since $f'(x) = \dfrac{1}{|x|\sqrt{x^2-1}} = \dfrac{1}{\sqrt{x^4-x^2}}$, $f''(x) = -\dfrac{x(2x^2-1)}{(x^4-x^2)^{3/2}}$. Since $f''(x) > 0$ if $-\infty < x < -1$, f is concave up on the interval $(-\infty, -1)$. Since $f''(x) < 0$ if $1 < x < \infty$, f is concave down on the interval $(1, \infty)$. f has no inflection points because there are no points in the domain of f at which f'' changes sign.

49. Since $(\arcsin(-x))' = -\dfrac{1}{\sqrt{1-x^2}} = (-\arcsin x)'$ for every x in the domain of the arcsine function, $\arcsin(-x) = -\arcsin x + C$ for some constant C. Evaluating both sides of this equation for a particular value of x (e.g., $x = 1$) shows that $C = 0$. Thus, $\arcsin(-x) = -\arcsin x$

51. Since $(\arccos x)' = -\dfrac{1}{\sqrt{1-x^2}} = (\pi/2 - \arcsin x)'$ for every x in the domain of the arcsine and the arccosine functions, $\arccos x = \pi/2 - \arcsin x + C$ for some constant C. Evaluating both sides of this equation for a particular value of x (e.g., $x = 0$) shows that $C = 0$. Thus, $\arccos x = \pi/2 - \arcsin x$.

53. $\left(\arcsin\left(\dfrac{x-1}{x+1}\right)\right)' = (2\arctan(\sqrt{x}))' = 1/(\sqrt{x}(x+1))$ so $\arcsin\left(\dfrac{x-1}{x+1}\right) = 2\arctan(\sqrt{x}) + C$, where C is a constant. Substituting $x = 0$ into this equation, we get $\arcsin(-1) = 2\arctan(0) + C \implies C = -\pi/2$.

55. (a) $f'(x) = 1/(1+x^2) = (\arctan x)' \implies f(x) = C + \arctan x$. Since $f(0) = \pi/4$, $f(x) = \pi/4 + \arctan x$ when $x < 1$.

(b) $\lim\limits_{x \to 1^-} f(x) = \pi/2$

(c) When $x > 1$, $x = 1/y$ where $0 < y < 1$. Therefore, $f(x) = f(1/y) = \arctan\left(\dfrac{1+1/y}{1-1/y}\right) = \arctan\left(\dfrac{1+y}{y-1}\right) = -\arctan\left(\dfrac{1+y}{1-y}\right) = -f(y)$. Thus, $\lim\limits_{x \to 1^+} f(x) = \lim\limits_{y \to 1^-} -f(y) = -\pi/2$.

57. (a) Let $x = \arctan u$ and $y = \arctan v$. Then, $\tan(\arctan u + \arctan v) = (u+v)/(1-uv) \implies \arctan u + \arctan v = \arctan\left(\dfrac{u+v}{1-uv}\right)$.

(b) The identity in part (a) is valid when $-\pi/2 < \arctan x + \arctan y < \pi/2$.

59. Let $f(x) = \arctan x - x/(1+x^2)$ and $g(x) = x - \arctan x$. Now,

$$f'(x) = \frac{1}{1+x^2} - \frac{1+x^2 - x(2x)}{(1+x^2)^2} = \frac{1+x^2 - (1-x^2)}{(1+x^2)^2} = \frac{2x^2}{(1+x^2)^2} \geq 0$$

for all x, and

$$g'(x) = 1 - \frac{1}{1+x^2} = \frac{1+x^2}{1+x^2} - \frac{1}{1+x^2} = \frac{x^2}{1+x^2} \geq 0$$

for all x. Since $f'(x) > 0$ and $g'(x) > 0$ for all x, and since $f(0) = g(0) = 0$, $f(x) \geq 0$ and $g(x) \geq 0$ for all $x \geq 0$. Thus, $x/(1+x^2) \leq \arctan x \leq x$ for all $x \geq 0$.

§3.5 Miscellaneous Derivatives and Antiderivatives

1. Using the product rule, $H'(z) = e^z(\cos z - \sin z)$.

3. Since $(\tan u)' = \sec^2 u$, $r'(\theta) = 3\sec^2(3\theta)$.

5. Using the product rule, $G'(w) = (2w + 3w^2)e^{3w}$.

7. Using the product rule and the chain rule,
$$W'(u) = 2ue^{u^2} \cdot \tan(u^{1/3}) + e^{u^2} \cdot \tfrac{1}{3}u^{-2/3}\sec^2(u^{1/3}) = \left(2u\tan(\sqrt[3]{u}) + \tfrac{1}{3}u^{-2/3}\sec^2(\sqrt[3]{u})\right)e^{u^2}.$$

9. Using the product rule, $g'(z) = 2ze^z + (z^2 + 3)e^z = (z^2 + 2z + 3)e^z$.

11. Using the chain rule, $f'(w) = 3 \cdot \left(-\sin(w^2)\right) \cdot 2w = -6w\sin(w^2)$

13. Using the product rule, $R'(w) = 3w^2\cos(4w) - 4w^3\sin(4w)$.

15. Using the product rule, $f'(x) = \sin x + x\cos x$.

17. Using the chain rule and the product rule, $h'(r) = e^{r^2 \sin r}(2r\sin r + r^2\cos r)$.

19. Using the chain rule, $f'(x) = e^{-2/x} \cdot (-2x^{-1})' = 2x^{-2}e^{-2/x} = \dfrac{2}{x^2 e^{2/x}}$.

21. Using the chain rule, $f'(x) = \dfrac{\cos(\sqrt{x})}{2\sqrt{x}}$

23. Using the chain rule, $f'(x) = 2\sec^2(2x + 3)$

25. Using the chain rule, $f'(x) = 4(e^{2x} - 3\ln x)^3 (2e^{2x} - 3/x)$

27. Using the chain rule and the quotient rule, $f'(x) = \dfrac{\frac{\sqrt{2+\cos x}}{x} + \frac{\ln x \sin x}{2\sqrt{2+\cos x}}}{2 + \cos x} = \dfrac{4 + 2\cos x + x\ln x \sin x}{2x(2 + \cos x)^{3/2}}$

29. Using the chain rule, $f'(x) = 2\cos(e^{3x}) \cdot 3e^{3x} = 6\cos(e^{3x})e^{3x}$

31. Using the chain rule, $f'(x) = 3\sin^2(e^{-4x})\cos(e^{-4x}) \cdot (-4e^{-4x}) = -12\sin^2(e^{-4x})\cos(e^{-4x})e^{-4x}$

33. Using the quotient rule, $f'(x) = \dfrac{\cos(x+1)(e^x + e^{-x}) - \sin(x+1)(e^x - e^{-x})}{(e^x + e^{-x})^2}$

35. Using the power rule, $f'(x) = e^{\pi} \cdot ex^{e-1} = e^{\pi+1}x^{e-1}$.

37. Using the product rule, $f'(x) = (\ln 2)2^x \ln x + 2^x/x$

39. Note that
$$f(x) = \log_2\left(\sqrt[3]{xe^x}\right) = \tfrac{1}{3}\log_2(xe^x) = \tfrac{1}{3}(\log_2 x + \log_2(e^x)) = \tfrac{1}{3}(\log_2 x + (\log_2 e)x) = \tfrac{1}{3}(\log_2 x + x/\ln 2).$$
Therefore, $f'(x) = \dfrac{1}{3}\left(\dfrac{1}{(\ln 2)x} + 1/\ln 2\right) = \dfrac{1+x}{3(\ln 2)x}$

41. Using the chain rule, $f'(x) = (\ln 4)4^{\sqrt{\ln x}} \cdot \left(\sqrt{\ln x}\right)' = (\ln 4)4^{\sqrt{\ln x}} \cdot \tfrac{1}{2}(\ln x)^{-1/2} \cdot x^{-1} = \dfrac{(\ln 4)4^{\sqrt{\ln x}}}{2x\sqrt{\ln x}}$.

43. Using the chain rule, $f'(x) = \dfrac{3e^{3x} + 2\sin x \cos x}{e^{3x} + \sin^2 x}$

45. Using the chain rule, $f'(x) = 2x\sec^2(x^2)e^{\tan(x^2)}$

SECTION 3.5 MISCELLANEOUS DERIVATIVES AND ANTIDERIVATIVES

47. Using the chain rule, $f'(x) = -3\cos(3x)\sin(3x)\left(4 + \cos^2(3x)\right)^{-1/2}$

49. Using the chain rule (or the product rule), $f'(x) = -6e^{3x}\cos\left(e^{3x}\right)\sin\left(e^{3x}\right)$

51. Using the product rule and the chain rule,
$$f'(x) = \ln(2 + \sin(3x)) + x(2 + \sin(3x))^{-1} \cdot 3\cos(3x) = \ln(2 + \sin(3x)) + \frac{3x\cos(3x)}{2 + \sin(3x))}.$$

53. Using the chain rule and the product rule, $f'(x) = 2\sin(x\cos x)\cos(x\cos x)(\cos x - x\sin x)$.

55. Using the chain rule and the quotient rule, $f'(x) = e^{x/\sin x} \cdot \left(\dfrac{x}{\sin x}\right)' = e^{x/\sin x} \cdot \dfrac{\sin x - x\cos x}{\sin^2 x}$.

57. Using the chain rule and the quotient rule,
$$f'(x) = -\sin\left(\frac{e^{2x}}{3+4x}\right) \cdot \left(\frac{e^{2x}}{3+4x}\right)' = -\sin\left(\frac{e^{2x}}{3+4x}\right) \cdot \frac{2e^{2x}(3+4x) - 4e^{2x}}{(3+4x)^2} = -\sin\left(\frac{e^{2x}}{3+4x}\right) \cdot \frac{(2+8x)e^{2x}}{(3+4x)^2}.$$

59. Using the chain rule and the quotient rule, $f'(x) = \dfrac{\frac{(2x+x^3)e^x}{1+e^x} - (2+3x^2)\ln(1+e^x)}{(2x+x^3)^2}$.

61. Use logarithmic differentiation: Let $g(x) = \ln(f(x)) = (\ln x) \cdot (\ln x) = (\ln x)^2$. Therefore,
$$g'(x) = \frac{2(\ln x)}{x} = \frac{f'(x)}{f(x)} \implies f'(x) = \frac{2(\ln x)}{x} \cdot x^{\ln x} = 2(\ln x)x^{\ln x - 1}.$$

63. Since $2x + 3 = (x^2 + 3x + 5)'$, $F(x) = \ln\left(x^2 + 3x + 5\right) + C$ is an antiderivative of f.

65. Since $\cos x = (2 + \sin x)'$, $F(x) = \ln(2 + \sin x) + C$ is an antiderivative of f.

67. $F(x) = \arcsin x + C$

69. $f(x) = \dfrac{3}{9 + x^2} = \dfrac{1}{3(1 + (x/3)^2)}$, so $F(x) = \arctan(x/3) + C$ is an antiderivative of f.

71. $f(x) = \dfrac{1}{\sqrt{1 - 4x^2}} = \dfrac{1}{\sqrt{1 - (2x)^2}}$, so $F(x) = \frac{1}{2}\arcsin(2x) + C$ is an antiderivative of f.

73. $f(x) = \dfrac{x}{\sqrt{1 - x^4}} = \dfrac{x}{\sqrt{1 - (x^2)^2}}$, so $F(x) = \frac{1}{2}\arcsin(x^2) + C$ is an antiderivative of f.

75. Let $g(x) = \sin x$. Then, $f(x) = \dfrac{f'(x)}{1 + (f(x))^2}$. Therefore, $F(x) = \arctan(\sin x) + C$ is an antiderivative of f.

77. Let $g(x) = \ln x$. Then, $f(x) = \dfrac{g'(x)}{1 + (g(x))^2}$ so $F(x) = \arctan(\ln x) + C$ is an antiderivative of f.

§3.6 Chapter summary

1. (a) By the chain rule, $h'(x) = f'(x^2)2x$, so $h'(0) = 0$.

   (b) If $f(u) = -u$, then $h(x) = f(x^2) = -x^2$ and $x = 0$ is a local maximum point. If $f(u) = u$, then $h(x) = f(x^2) = x^2$ and $x = 0$ is a local minimum point.

   (c) No, because $h'(x) = 2xf'(x^2)$ must change sign at $x = 0$. Thus, h cannot have a terrace point at $x = 0$.

3. (a) $(uv)'(3) = u'(3) \cdot v(3) + u(3) \cdot v'(3) \approx 1.5$
   $$(u/v)'(3) = \frac{u'(3) \cdot v(3) - u(3) \cdot v'(3)}{(v(3))^2} \approx 1.5$$
   $(u \circ v)'(3) = u'(v(3)) \cdot v'(3) \approx -1.8$

   (b) uv has a stationary point at $x = 2$ because $(uv)'(2) = u'(2) \cdot v(2) + u(2) \cdot v'(2) = 0$. This stationary point is a local minimum point because the sign of $(uv)'(x)$ changes from negative to positive at $x = 2$.

   (c) $v \circ u$ has a stationary point at $x = 2$ because $(v \circ u)'(2) = v'(u(2)) \cdot u'(2) = 0$. This stationary point is a local minimum point because the sign of $(v \circ u)'(x)$ changes from negative to positive at $x = 2$.

   (d) $(u \circ v)'(x) = u'(v(x))v'(x)$ so $u \circ v$ has stationary point at x if $v'(x) = 0$ or if $u'(v(x)) = 0$. Thus, $u \circ v$ has stationary points at $x \approx 0.02$, $x \approx 0.7$, $x \approx 1.7$, $x \approx 2.8$, and $x \approx 3.7$.

   (e) The function $u \circ u$ has a stationary point at each value of x where $(u \circ u)'(x) = 2u'(u(x)) \cdot u'(x) = 0$. This implies that $u \circ u$ has a stationary point at x if $u'(x) = 0$ (i.e., if x is a stationary point of u) or if $u(x) = s$, where s is a stationary point of u. Thus, u has stationary points at $x = 0.59$, $x = 2$, and $x = 3.41$.
   Symbolically, $(u \circ u)(x) = x^4 - 8x^3 + 20x^2 - 16x + 4$. Therefore, $(u \circ u)'(x) = 4x^3 - 24x^2 + 40x - 16$ so $u \circ u$ has stationary points at $x = 2 - \sqrt{2}$, $x = 2$, and $x = 2 + \sqrt{2}$.

5. Differentiating $x + y = k$ implicitly gives $1 + dy/dx = 0$, so $dy/dx = -1$. The DE implies that the solution functions are lines of slope -1.

7. (a) The curves have k-values $\pm 1, \pm 2, \pm 3$.

   (b) Implicit differentiation of $xy = k$ gives $y + xy' = 0$, so $y' = -y/x$.

   (c) If $y = k/x$, then $y' = -k/x^2 = -y/x$.

   (d) The DE means that the line from the origin to (x, y) and the tangent line at (x, y) have opposite slopes. Equivalently, these two lines and the x-axis form an isosceles triangle.

9.
x	$f(x)$	$f'(x)$	$g(x)$	$g'(x)$	$h(x)$	$h'(x)$	$j(x)$	$j'(x)$
-1	3	2	1	1/10	0	$-1/2$	3	2.3
0	0	1/2	-1	1	3	2	0	$-1/2$
1	0	-5	-1	1	3	2	0	5

11. If C is a real number, then $x(\ln x - 1) + C$ is an antiderivative of the natural logarithm function $\ln x$. Since $\log_b x = \ln x / \ln b$, $x(\ln x - 1)/\ln b + C = x \left(\log_b x - \frac{1}{\ln b} \right) + C$ is an antiderivative of $\log_b x$ for any real number C.

13. $h'(x) = e^x f'(e^x)$ so $h'(\ln 2) = e^{\ln 2} f'(e^{\ln 2}) = 2f'(2) = -6$. Since $h(\ln 2) = f(e^{\ln 2}) = f(2) = 4$, $y = -6(x - \ln 2) + 4$ is an equation of the tangent line.

Section 3.6 Chapter Summary

15. No — $k'(x) = \dfrac{f'(x) \cdot (1+e^x) - f(x) \cdot e^x}{(1+e^x)^2} \implies k'(0) = \dfrac{2f'(0) - f(0)}{4} = -\dfrac{5}{4} > -2$.

17. No, $G'(u) = 1 + \ln u + \frac{u}{u} = 2 + \ln u \neq g(u)$.

19. No. $y'(x) = 10x^4 \neq 10y/x = 10 \cdot (2x^5)/x = 20x^4$.

21. $\bigl(f(h(x))\bigr)' = f'(h(x)) \cdot h'(x) = g(h(x)) \cdot 2x = 2xg(x^2)$

23. $\dfrac{d}{dx}(g(x))^n = n(g(x))^{n-1} g'(x)$

25. $\dfrac{d}{dx} \ln(g(x)) = \dfrac{g'(x)}{g(x)}$

27. $\dfrac{d}{dx} \tan(g(x)) = g'(x) \sec^2(g(x))$

29. $\dfrac{d}{dx} g(e^x) = e^x g'(e^x)$

31. $\dfrac{d}{dx} g(\sin x) = \cos x\, g'(\sin x)$

33. (a) $N(t) = A\left(25 + te^{-t/20}\right)$, where A is the constant of proportionality. Therefore, $N'(t) = Ae^{-t/20}(1 - t/20)$, so $N'(t) > 0$ if $0 \le t < 20$, and $N'(t) < 0$ if $t > 20$. Thus, the number of bacteria is a minimum at $t = 0$.

   (b) Using the computations done for part (a), we see that the number of bacteria in the culture is a maximum at $t = 20$.

   (c) $N''(t) = Ae^{-t/20}(t-40)/400$, so the rate of change of N' has a minimum at $t = 40$.

35. (a) $(f(x))^2 = (f(x) \cdot f(x))$, so $\dfrac{d}{dx}(f(x))^2 = f'(x)f(x) + f(x)f'(x) = 2f(x)f'(x)$.

    (b) $(h(x))' = \dfrac{d}{dx}(f(x))^3 = 3(f(x))^2 f'(x)$.

37. $\lim\limits_{x \to \pi} \dfrac{x \cos x + \pi}{x - \pi} = f'(a)$, where $f(x) = x \cos x$ and $a = \pi$. Now, $f'(x) = \cos x - x \sin x$ so $f'(\pi) = -1$. Therefore, $\lim\limits_{x \to \pi} \dfrac{x \cos x + \pi}{x - \pi} = -1$.

39. Since the expression $(f(1))^2 + (g(1))^2$ is a sum of squares, it is nonnegative. Furthermore, the value of this expression is zero if and only if $f(1) = g(1) = 0$. However, if $f(1) = g(1) = 0$, $h'(1) = f'(1)g(1) + f(1)g'(1)$ would be zero. Since $h'(1) \neq 0$, we conclude that $f(1) = g(1) = 0$ cannot be true.

41. Using the chain rule, $h'(w) = (w \ln w)' \cdot f'(w \ln w) = (1 + \ln w)\sqrt{1 + (w \ln w)^3}$. Thus, $h'(\pi) = (1 + \ln \pi)\sqrt{1 + (\pi \ln \pi)^3} \approx 14.783$.

43. No — $X'(s) = -s^2 e^{-s} = -Y(s)$.

45. No — $X'(z) = 2e^z \cos z \neq e^z \sin z$.

47. No. $F'(x) = e^{\sin x}(\cos^2 x - \sin x) \neq f(x)$.

49. $\arctan(\tan \theta) = \theta$ if $-\pi/2 < \theta < \pi/2$. This is because the range of the arctangent function is the interval $(-\pi/2, \pi/2)$.

51. The domain of the arcsine function is the interval $[-1, 1]$. Thus, $\arcsin t$ is defined if $-1 \le t \le 1$.

53. (a) $f(0) = f(0+0) = f(0) \cdot f(0) = (f(0))^2$ so either $f(0) = 0$ or $f(0) = 1$. However, by hypothesis, $f(0) \neq 0$ so we may conclude that $f(0) = 1$.

(b) Since $1 = f(0) = f(x + (-x)) = f(x)f(-x)$, $f(x) \neq 0$ for all x.

(c) $f'(x) = \lim_{h \to 0} \dfrac{f(x+h) - f(x)}{h} = \lim_{h \to 0} \dfrac{f(x)f(h) - f(x)}{h} = f(x) \cdot \lim_{h \to 0} \dfrac{f(h) - 1}{h} = f(x) \cdot \lim_{h \to 0} \dfrac{f(h) - f(0)}{h} = f(x) \cdot f'(0) = f(x)$

(d) Since $g(x) \neq 0$ for all x, $k(x)$ is defined for all x. Now,

$$k'(x) = \dfrac{f'(x)g(x) - f(x)g'(x)}{(g(x))^2} = \dfrac{f(x)g(x) - f(x)g(x)}{(g(x))^2} = 0$$

so k is a constant function.

(e) Since $k(0) = 1$, $k(x) = 1$ for all x (i.e., $f(x) = g(x)$). Therefore, since the exponential function e^x satisfies conditions (i), (iii), and (iv), it is the *only* function that satisfies these properties.

§4.1 Slope Fields; More Differential Equation Models

1. (a) The solution curves are "parallel" to each other in the sense that they differ from each other only in their *horizontal* position. Thus, e.g., all the curves have the same slope where $y = 2$.

   (b) It *does* appear that each of the five "upper" curves has the same slope when $y = 3$. Carefully draw a tangent line to any one of the curves at the appropriate point; measure its slope. The result should be 3 (or very close to 3).

   The answer *could* have been predicted in advance. The fact that each curve is a solution to the DE $y' = y$ means precisely that when $y = 3$, $y' = 3$, too.

   (c) At the level $y = -4$, each curve has slope -4. Again, this is exactly what the DE predicts.

   (d) All curves appear to be very nearly *horizontal* near $y = 0$. The only solution curve that actually touches the line $y = 0$ is the solution curve $y = 0$ itself. Appropriately, this curve has slope 0 everywhere.

3. Since $f(1) = 2$, the point $(1, 2) = (t, y(t))$ is on the solution curve. From the DE, we find that $y'(1) = 1^2 \cdot y(1) + 1 = 3$ so the tangent line has slope 3. Since the tangent line passes through the point $(1, 2)$, it is described by the equation $y = 3t - 1$.

5. The slope of the tangent line at $(-2, 1)$ is $y'(-2) = (-2)^2 \cdot y(-2) = 4$. Thus, $y = 4(t + 2) + 1 = 4t + 9$ is an equation of the tangent line.

7. This is the slope field of DE (vii), $y' = \cos y$. Observe that the slope is the same at each x-value — the slope at each grid point depends only on the value of y.

9. This is the slope field of DE (iii), $y' = ty$.

11. This is the slope field of DE (viii), $y' = \sin t$.

13. This is the slope field of DE (x), $y' = y(1 - y)$.

15. All ticks at the same vertical position are parallel.

17. (b) If $y(t) = \sqrt{t^2 + C}$, then $y'(t) = t/\sqrt{t^2 + C} = t/y$. Similarly, if $y(t) = -\sqrt{t^2 + C}$, then $y'(t) = -t/\sqrt{t^2 + C} = t/y$.

    (c) $y(t) = \sqrt{t^2 + 1}$

    (d) $y(t) = \pm\sqrt{t^2 - 1}$

19. (b) If $y(t) = e^{-t} + Ce^{-2t}$, then $y'(t) = -e^{-t} - 2Ce^{-2t} = e^{-t} - 2\left(e^{-t} + Ce^{-2t}\right) = e^{-t} - 2y$.

    (c) $y(t) = e^{-t} - e^{-2(t+1)}$

    (d) $y(t) = e^{-t} - e^{2-2t} - e^{1-2t}$

    (e) $y(t) = e^{-t} + e^{2-2t} - e^{1-2t}$

21. (a) $y(t) = t - 1 + 2e^{-t}$

    (b) $y(t) = t - 1 + 4e^{-t}$

    (c) $y(t) = t - 1$

    (d) If $y(t) = t - 1 + Ce^{-t}$, then $y'(t) = 1 - Ce^{-t} = t - (t - 1 + Ce^{-t}) = t - y(t)$.

    (e) $y' = t - y \implies y''(t) = 1 - y'(t) = 1 - (t - y(t)) = 1 - t + y(t)$. This implies that the solution curve in part (a) is concave up, since $y''(t) > 0$; the solution curve in part (b) is concave down since $y''(t) < 0$, and the solution curve in part (c) is linear since $y''(t) = 0$.

23. (a) Since the outside temperature is $-10°$ C and the coffee is initially $90°$ C, the temperature $y(t)$ of the coffee at time t is $y(t) = -10 + 100e^{kt}$. For the foam cup, $k = -0.05$, so the temperature reaches $70°$ C at time $t = -\ln(80/100)/0.05 \approx 4.46$ minutes after leaving the store. For the cardboard cup, $k = -0.08$, so the temperature reaches $70°$ C after $t = -\ln(80/100)/0.08 \approx 2.79$ minutes.

After $t = 5$ minutes, the coffee in the foam cup is $-10 + 100e^{-0.25} \approx 68°$ C while the coffee in the paper cup is $-10 + 100e^{-0.4} \approx 57°$ C.

(b) In the store, the temperature of the coffee at time t is $y(t) = 25 + 65e^{kt}$. Therefore, the coffee reaches $70°$ C after $t = \ln(45/65)/k$ minutes. This implies that Boris's coffee reaches $70°$ C in approximately 7.4 minutes and Natasha's coffee reaches this temperature in approximately 4.6 minutes.

After $t = 5$ minutes, Boris's coffee is approximately $76°$ C and Natasha's coffee is approximately $69°$ C.

(c) If the coffee is to be at least $70°$ C after 5 minutes outdoors, k must be chosen so that $70 \leq -10 + 100e^{5k}$. This implies that $k \geq \ln(80/100)/5 \approx -0.0446$.

25. (a) $P(t)$ is an increasing function, so $P'(t)$ must never be negative. The factor $M - P(t)$ causes the rate of learning to decrease as the value of the performance function approaches the maximum. Thus, the rate of learning is large when $M - P(t)$ is large and approaches zero as $P(t)$ approaches M. The value of P never exceeds M because $P' = 0$ when $P = M$.

(b) If $P(t) = M - Ae^{-kt}$, then $P'(t) = Ake^{-kt} = kAe^{-kt} = k(M - P)$, as desired.

(c) From the previous part, solutions are of the form $P(t) = M - Ae^{-0.05t}$. Since $P(0) = 0.1M$, it follows that $A = 0.9M$. Therefore, $P(t) = M - 0.9Me^{-0.05t}$.

We want t such that $P(t) = 0.9M$, so we solve the equation $M - 0.9Me^{-0.05t} = 0.9M$ for t. The solution is $t = 20 \ln 9 \approx 44$ hours.

27. (a) The DE $v' = g - kv$ can be rewritten in the form $v' = (-k)(v - (-g/k))$ which has the same form as the DE describing Newton's law of cooling. Thus, the solution of the DE is
$$v(t) = \frac{g}{k} - \left(\frac{g}{k} - v_0\right)e^{-kt}$$

(b) Since $y'(t) = v(t)$, $y(t) = \frac{g}{k}t + \left(\frac{g}{k^2} - \frac{v_0}{k}\right)e^{-kt} + \left(y_0 - \frac{g}{k^2} + \frac{v_0}{k}\right)$.

29. Observe that the graph of $y = f(x)$ is decreasing when $1 < x < 2$ and increasing when $2 < x < 5$. Thus, y' must be negative on $(1, 2)$ and positive on $(2, 5)$. Also, observe that $y > x$ on $(1, 2)$ and $y < x$ on $(2, 5)$. Furthermore, $y < x^2$ on $(2, 5)$.

The differential equation (a) cannot be the correct answer since the expression $(y - x)/x$ is negative on the interval $(2, 5)$. Similarly, the differential equation (c) cannot be the correct answer because the expression $(x^2 - y)/x$ is not negative over the entire interval $(1, 2)$. On the other hand, the expression $(x - y)/y$ is negative on the interval $(1, 2)$ and positive on the interval $(2, 5)$. Therefore, the correct differential equation is (b).

31. For each value of y, the slopes do not depend on x (i.e., all ticks at the same vertical position are parallel).

33. (b) Since the DE does not depend explicitly on t, the solutions for different initial conditions can be obtained by horizontal shifts. Thus, since $y(t) = (t + 1)e^t$ is the solution of the DE with the initial condition $y(0) = 1$, $y(t) = (t + 2)e^{t+1}$ is the solution of the DE with the initial condition $y(0) = 2e$.

§4.2 More on Limits: Limits Involving Infinity and l'Hôpital's Rule

1. Note that $p(x) = (x+1)(2-x) = 2 + x - x^2 = x^2(2/x^2 + 1/x - 1)$.

   (a) Since $\lim_{x\to\infty} (2/x^2 + 1/x - 1) = -1$ and $\lim_{x\to\infty} x^2 = \infty$, $\lim_{x\to\infty} p(x) = -\infty$.

   (b) Since $\lim_{x\to-\infty} (2/x^2 + 1/x - 1) = -1$ and $\lim_{x\to-\infty} x^2 = \infty$, $\lim_{x\to-\infty} p(x) = -\infty$.

3. $\lim_{x\to 2} \dfrac{f(x)}{g(x)}$ does not exist because $\lim_{x\to 2^-} \dfrac{f(x)}{g(x)} = \infty$ but $\lim_{x\to 2^+} \dfrac{f(x)}{g(x)} = -\infty$.

5. Near $x = -2$, $f(x) = -x - 2 = -(x+2)$. Therefore, $\lim_{x\to -2} \dfrac{x+2}{f(x)} = \lim_{x\to -2} \dfrac{x+2}{-(x+2)} = -1$.
   [NOTE: l'Hôpital's rule could also be used to obtain this result.]

7. $\lim_{x\to 1} \dfrac{f(x)+2}{g(x)+1}$ does not exist because $\lim_{x\to 1^-} \dfrac{f(x)+2}{g(x)+1} = \lim_{x\to 1^-} \dfrac{-2x+2}{-x+1} = 2$ but $\lim_{x\to 1^+} \dfrac{f(x)+2}{g(x)+1} = \lim_{x\to 1^+} \dfrac{x-3}{x-2} = 1$.

9. $\lim_{x\to\infty} \dfrac{3}{x^2} = 0$

11. $\lim_{t\to\infty} \dfrac{2t+3}{5-4t} = -\dfrac{1}{2}$.

13. $\lim_{x\to\infty} \dfrac{x^2+1}{x} = \infty$

15. $\lim_{x\to\infty} \dfrac{\sin x}{x} = 0$

17. $\lim_{x\to\infty} \dfrac{2^x}{x^2} = \infty$

19. $\lim_{x\to\infty} \dfrac{\ln x}{x^{2/3}} = \lim_{x\to\infty} \dfrac{x^{-1}}{2x^{-1/3}/3} = \lim_{x\to\infty} \dfrac{3}{2x^{2/3}} = 0$.

21. $\lim_{x\to 0} \sin(\sin x) = 0$

23. $\lim_{x\to 0} \dfrac{\tan x}{x} = \lim_{x\to 0} \sec^2 x = 1$ so $\lim_{x\to 0} \cos\left(\dfrac{\tan x}{x}\right) = \cos 1$

25. $\lim_{x\to 1} \dfrac{x^3+x-2}{x^2-3x+2} = \lim_{x\to 1} \dfrac{3x^2+1}{2x-3} = -4$

27. $\lim_{x\to 0} \dfrac{1-\cos x}{\sin(2x)} = \lim_{x\to 0} \dfrac{\sin x}{2\cos(2x)} = 0$

29. $\lim_{x\to\infty} \dfrac{e^x}{x^2+x} = \lim_{x\to\infty} \dfrac{e^x}{2x+1} = \lim_{x\to\infty} \dfrac{e^x}{2} = \infty$

31. Since $\lim_{x\to 1} f(x) = \lim_{x\to 1} x^2 - 1 = 0$, l'Hôpital's rule can be used to evaluate the desired limit. Thus,
    $\lim_{x\to 1} \dfrac{f(x)}{x^2-1} = \lim_{x\to 1} \dfrac{f'(x)}{2x} = \dfrac{f'(1)}{2}$. It appears from the graph that $f'(1) \approx 3/2$, so $\lim_{x\to 1} \dfrac{f(x)}{x^2-1} \approx \dfrac{3}{4}$.

33. $\lim_{x\to 4} \dfrac{f(x)}{(x-4)^2} = \infty$ since $\lim_{x\to 4} f(x) \approx 3.6$ and since $\lim_{x\to 4} (x-4)^{-2} = \infty$.

35. Since $\lim_{x\to 1} f(x-3) = 0$ and $\lim_{x\to 1} f(x+3) = 3.6$, $\lim_{x\to 1} \dfrac{f(x-3)}{f(x+3)} = 0$.

37. No. Since $\lim_{x\to \pi/2^-} \tan x = \infty$ and $\lim_{x\to \pi/2^+} \tan x = -\infty$, $\lim_{x\to \pi/2} \tan x$ does not exist. Therefore, l'Hôpital's rule cannot be used. Moreover, the limits $\lim_{x\to \pi/2^\pm} \dfrac{\tan x}{x - \pi/2}$ have the form $\mp\infty/0$ so l'Hôpital's rule cannot be used to evaluate these limits either.
[NOTE: L'Hôpital's rule can be used only for limits of the form $0/0$ or ∞/∞.]

39. No. The limit $\lim_{x\to \pi/2^-} \dfrac{\tan x}{x - \pi/2}$ has the form $\infty/0$ so l'Hôpital's rule cannot be used to evaluate this limit.
[NOTE: L'Hôpital's rule can be used only for limits of the form $0/0$ or ∞/∞.]

41. Since f is a non-constant periodic function, the values of $f(x)$ do not approach a single number L as $x \to \infty$.

43. There are many ways to draw such a graph. Any suitable graph, though, should have a horizontal asymptote at $y = -3$ (to the right) and another horizontal asymptote at $y = 3$ (to the left). A simple way to satisfy condition (iii) is to have the graph pass through $(2, -2)$.

45. (a) Yes. The line $y = -1$ is a horizontal asymptote of h since $\lim_{x\to\infty} h(x) = -1$.

(b) No. If h were a rational function with a vertical asymptote at $x = 3$ and the property $\lim_{x\to\infty} h(x) = -1$, then h would have to be of the form
$$h(x) = \dfrac{-x^n + \cdots}{(x-3)(x^{n-1} + \cdots)}$$
for some positive integer n. But then $\lim_{x\to -\infty} h(x) = -1$ not ∞.

47. p and q can be any polynomials such that the coefficient of the highest power of x in each polynomial is positive, and the degree of p is greater than the degree of q. For example, the polynomials $p(x) = x^2 - 2x + 3$ and $q(x) = x + 1$ have the desired properties: $\lim_{x\to\infty} p(x) = \lim_{x\to\infty} q(x) = \lim_{x\to\infty} p(x)/q(x)$.

49. p and q can be any polynomials such that the degree of p is less than the degree of q. For example, the polynomials $p(x) = x + 1$ and $q(x) = x^2 + 2$ have the desired properities: $\lim_{x\to\infty} p(x) = \lim_{x\to\infty} q(x) = \infty$, but $\lim_{x\to\infty} p(x)/q(x) = 0$.

51. (a) If $a = -21$, then $\lim_{x\to 3} f(x) = \dfrac{13}{4}$.

(b) $\lim_{x\to\infty} f(x) = 2$ for every value of a.

53. g is **not** continuous at $x = 0$ because $\lim_{x\to 0^-} f(x) \neq \lim_{x\to 0^+} f(x)$.

55. $\lim_{x\to\infty} e^{-x} \ln x = \lim_{x\to\infty} \left(\dfrac{\ln x}{e^x}\right) = \lim_{x\to\infty} \left(\dfrac{\frac{1}{x}}{e^x}\right) = \lim_{x\to\infty} \left(\dfrac{1}{xe^x}\right) = 0$ [Denominator blows up!]

57. $\lim_{x\to 8} \left(\dfrac{x-8}{\sqrt[3]{x}-2}\right) = \lim_{x\to 8} \left(\dfrac{1}{\frac{1}{3}x^{-2/3}}\right) = \dfrac{1}{\frac{1}{3}\cdot 8^{-2/3}} = 3 \cdot 8^{2/3} = 3 \cdot 4 = 12$

59. $\lim_{x\to 0}\left(\dfrac{\sin x}{x - \sin x}\right) = \lim_{x\to 0}\left(\dfrac{\cos x}{1 - \cos x}\right) = \infty$ [NOTE: numerator $\to$ 1 and denominator $\to$ 0 from above.]

SECTION 4.2 MORE ON LIMITS: LIMITS INVOLVING INFINITY AND L'HÔPITAL'S RULE

61. $\lim\limits_{x \to 0} \dfrac{e^x - 1}{x} = \lim\limits_{x \to 0} \dfrac{e^x}{1} = 1$

63. $\lim\limits_{x \to 0} \dfrac{e^x - e^{-x}}{x} = \lim\limits_{x \to 0} \left(e^x + e^{-x} \right) = 2$

65. $\lim\limits_{x \to 0} \dfrac{1 - \cos^2 x}{x^2} = \lim\limits_{x \to 0} \dfrac{2 \cos x \sin x}{2x} = \lim\limits_{x \to 0} \dfrac{2 \cos^2 x - 2 \sin^2 x}{2} = 1$

67. $\lim\limits_{x \to 1} \dfrac{\ln x}{x^2 - x} = \lim\limits_{x \to 1} \dfrac{1}{2x^2 - x} = 1$

69. $\lim\limits_{x \to 1} \dfrac{\sin(\pi x)}{x^2 - 1} = \lim\limits_{x \to 1} \dfrac{\pi \cos(\pi x)}{2x} = -\dfrac{\pi}{2}$

71. $\lim\limits_{x \to 1} \dfrac{\cos^3(\pi x/2)}{\sin(\pi x)} = \lim\limits_{x \to 1} \dfrac{-3 \cos^2(\pi x/2) \sin(\pi x/2)}{2 \cos(\pi x)} = 0$

73. $\lim\limits_{x \to 0^+} x^2 \ln x = \lim\limits_{x \to 0^+} \dfrac{\ln x}{x^{-2}} = \lim\limits_{x \to 0^+} -\dfrac{x^2}{2} = 0$

75. $\lim\limits_{x \to 0} x^2 \ln(\cos x) = 0$ since $\lim\limits_{x \to 0} x^2 = 0$ and $\lim\limits_{x \to 0} \ln(\cos x) = 0$.

77. $\lim\limits_{w \to 0^+} w(\ln w)^2 = \lim\limits_{w \to 0^+} \dfrac{(\ln w)^2}{1/w} = \lim\limits_{w \to 0^+} -\dfrac{2 \ln w}{1/w} = \lim\limits_{w \to 0^+} 2w = 0.$

79. $\lim\limits_{x \to \infty} x \left(\dfrac{\pi}{2} - \arctan x \right) = \lim\limits_{x \to \infty} \dfrac{\pi/2 - \arctan x}{x^{-1}} = \lim\limits_{x \to \infty} \dfrac{x^2}{1 + x^2} = 1$

81. $\lim\limits_{x \to 0} \left(\dfrac{1}{\sin x} - \dfrac{1}{x} \right) = \lim\limits_{x \to 0} \dfrac{x - \sin x}{x \sin x} = \lim\limits_{x \to 0} \dfrac{1 - \cos x}{\sin x + x \cos x} = \lim\limits_{x \to 0} \dfrac{\sin x}{2 \cos x - x \sin x} = 0$

83. Using l'Hôpital's rule, $\lim\limits_{x \to 1} \dfrac{(f(x))^2 - 4}{x^2 - 1} = \lim\limits_{x \to 1} \dfrac{f(x) f'(x)}{x} = 6.$

85. $\lim\limits_{x \to 0} \dfrac{x f(x)}{(e^x - 1) g(x)} = \lim\limits_{x \to 0} \dfrac{f(x) + x f'(x)}{e^x g(x) + (e^x - 1) g'(x)} = \dfrac{f(0)}{g(0)}$

87. $\lim\limits_{x \to 0^+} 2x \ln x = 0$ so $\lim\limits_{x \to 0^+} x^{2x} = e^0 = 1$

89. $\lim\limits_{x \to 0} \dfrac{\ln(1 + x)}{x} = \lim\limits_{x \to 0} \dfrac{1}{1 + x} = 1$ so $\lim\limits_{x \to \infty} (1 + x)^{1/x} = e^1 = e.$

91. $\lim\limits_{x \to 1} (\ln x)^{\sin x} = 0$ since $\lim\limits_{x \to 1} \ln x = 0$ and $\lim\limits_{x \to 1} \sin x = \sin 1 \approx 0.84147.$

93. $\lim\limits_{x \to 0} \dfrac{f(x)}{x} = \lim\limits_{x \to 0} \dfrac{f(x) - f(0)}{x - 0}$. The limit on the right is the definition of $f'(0)$.

95. $f'(1)$ does not exist because f is not continuous at $x = 1$. Here are the details:
$f'(1) = \lim\limits_{h \to 0} \dfrac{f(1 + h) - f(1)}{h}$. However, the two-sided limit in the definition of $f'(1)$ does not exist
because $\lim\limits_{h \to 0^+} \dfrac{f(1+h) - f(1)}{h} = \lim\limits_{h \to 0^+} \dfrac{(1+h)^2 + 3 - 5}{h} = \lim\limits_{h \to 0^+} \dfrac{h^2 + 2h - 1}{h} = -\infty$ and
$\lim\limits_{h \to 0^-} \dfrac{f(1+h) - f(1)}{h} = \lim\limits_{h \to 0^-} \dfrac{2(1+h) + 2 - 5}{h} = \lim\limits_{h \to 0^-} \dfrac{2h - 1}{h} = \infty.$

§4.3 Optimization

1. A function achieves its minimum value over a closed interval at an endpoint of the interval or at a stationary point within the interval. Since $f'(1) = 0$, $f(x)$ could attain its minimum value at $x = -5, x = 1$, or $x = 5$.

3. Yes — the function h fails to be differentiable at $x = 1$ so it has a critical point there.

5. (a) We can write $g(x) = 5x^2 - 8x + 4$, so $g'(x) = 10x - 8$. The only stationary point occurs where $g'(x) = 10x - 8 = 0$, i.e., at $x = 4/5$. Also, $g(4/5) = 4/5$, so $4/5$ is also the minimum value of g.

   (b) The previous part shows $g(x)$ is least if $x = 4/5$. This means that the square of the distance (and hence the distance itself) is least if $x = 4/5$, just as we found in Example 3. This minimum distance, moreover, is $\sqrt{4/5} = 2/\sqrt{5} \approx 0.8944$.

7. (a) The sketch should suggest that P has coordinates somewhere near $(0.7, 0.5)$.

   (b) The function $f(x) = x^2 + (1 - x^2)^2$ gives the square of the distance from the origin to the point $(x, 1 - x^2)$ on the parabola. Minimizing f in the usual way shows that the minimum distance occurs if $x = \sqrt{2}/2$ and $y = 1/2$—i.e., P is the point $(\sqrt{2}/2, 1/2)$.

   (c) The line from the origin to $P(\sqrt{2}/2, 1/2)$ has slope $1/\sqrt{2}$. At $x = \sqrt{2}/2$, the parabola $y = 1 - x^2$ has slope $-\sqrt{2}$. This means that the two lines are perpendicular, as claimed.

9. The equations $P = xy$ and $x + y = 10$ imply that $P(x) = x(10 - x) = 10x - x^2$; the exercise is meaningful for $0 \le x \le 10$. Now $P'(x) = 10 - 2x = 0$ if $x = 5$, and $P(5) = 25$.

11. The constraint $x + y = 4$ implies that $y = 4 - x$. Thus, we wish to find the minimum value of $f(x) = x^3 + (4 - x)$ if $x \ge 0$. Since $f'(x) = 3x^2 - 1$, the only critical point of f in the interval $[0, \infty)$ is at $x_* = 1/\sqrt{3} = \sqrt{3}/3$. Since $f''(x_*) = 6x_* = 2\sqrt{3} > 0$, this value of x is a local minimum point of f. It follows that the minimum value of f over the interval $[0, \infty)$ is
$f(x_*) = x_*^3 + 4 - x_* = (\sqrt{3}/3)^3 + 4 - \sqrt{3}/3 = 4 - 2\sqrt{3}/9 \approx 3.6151$.

13. If $H(w) = 4w^5 - 5w^6$, $H'(w) = 20w^4 - 30w^5 = 10w^4(2 - 3w)$. Thus, $H'(w) > 0$ on the interval $(-\infty, 2/3)$ and $H'(w) < 0$ on the interval $(2/3, \infty)$. This implies that H is increasing everywhere to the left of $w = 2/3$ and decreasing everywhere to the right of $w = 2/3$. Therefore, $w = 2/3$ is the global maximum point of $H(w)$.

15. If the base of the rectangle is on the x-axis and has corners at $(-a, 0)$ and $(a, 0)$, the area of the inscribed rectangle is $A(a) = 2a \cdot \sqrt{3}(1 - a) = 2\sqrt{3}a(1 - a)$. Now, $A'(a) = 2\sqrt{3}(1 - 2a) = 0$ if $a = 1/2$. Since $A(0) = A(1) = 0$ and $A(1/2) = \sqrt{3}/2$, $a = 1/2$ corresponds to the rectange with largest area. This rectangle has corners at the points $(\pm 1/2, 0), (\pm 1/2, \sqrt{3}/2)$; it occupies 50% of the area of the triangle.

17. Suppose that the side of the triangle in the first quadrant is tangent to the parabola at the point $x = a > 0$. Then the tangent line is described by the equation $y = -2a(x - a) + (1 - a^2)$. The x-intercept of this tangent line is the point $\big((1 + a^2)/2a, 0\big)$; its y-intercept is the point $(0, 1 + a^2)$. By symmetry, the area of the triangle corresponding to this tangent line is $A(a) = \dfrac{1}{2} \cdot \dfrac{1 + a^2}{a} \cdot (1 + a^2) = \dfrac{(1 + a^2)^2}{2a}$.

    Since we wish to find the value of a that minimizes $A(a)$, we solve for the stationary points of A': $A'(a) = 0 \implies a = \pm\sqrt{3}/3$. Since $A'(x) < 0$ if $0 < x < \sqrt{3}/3$ and $A'(x) > 0$ if $\sqrt{3}/3 < x < 1$, $a = \sqrt{3}/3$ corresponds to the triangle with smallest possible area. Its area is $A(\sqrt{3}/3) = 8\sqrt{3}/9$ square units.

19. (a) The trajectory formula gives $y = x - 9.8x^2/900$. The peak occurs where $dy/dx = 1 - 9.8x/450 = 0$, i.e., at $x \approx 45.92$ meters. If $x \approx 45.92$, then $y \approx 22.96$ meters. The ball lands where $y = 0$, i.e., at $x \approx 91.84$ meters.

    (b) The trajectory is a parabola, intersecting the x-axis at $x = 0$ and $x = 91.84$; the peak occurs at $(45.92, 22.96)$.

SECTION 4.3 OPTIMIZATION

21. (a) Set the trajectory equation to zero and solve for x. There are two solutions—$x = 0$ and $x = R$, where R is the range.

    (b) Let $R(m) = \dfrac{2v_0^2}{g} \dfrac{m}{1+m^2}$; let's maximize this for $m > 0$. Well, $R'(m) = \dfrac{2v_0^2}{g} \dfrac{1-m^2}{(1+m^2)^2}$; thus $R'(m) = 0$ for $m = 1$. Thus the range is maximum if the initial slope is 1, or, equivalently, the initial angle is $\pi/4$.

23. (a) Maximizing $A = xy$ subject to $2hx + 2vy = b$ leads to $hx = vy = b/4$, so $x = b/(4h)$, $y = b/(4v)$, and the maximum possible area $xy = b^2/(16hv)$ square feet.

    (b) Minimizing $C = 2hx + 2vy$ subject to $xy = a$ leads to $hx = vy$. Combining this with $a = xy$ gives $x = \sqrt{av/h}$ and $y = \sqrt{ah/v}$. Thus the minimum possible cost is $2hx + 2vy = 4\sqrt{avh}$.

    (c) The two parts are consistent—both say that the best scheme is to spend half the money on east-west fence and half on north-south fence.

25. The volume of the can is $V = \pi r^2 h$ and the surface area is $A = 2\pi rh + 2\pi r^2$. Since the can must hold 168 cm^3, $168 = \pi r^2 h$ or $h = 168/\pi r^2$. This allows us to express A as a function of r:

$$A(r) = 2\pi r \left(\dfrac{168}{\pi r^2}\right) + 2\pi r^2 = \dfrac{336}{r} + 2\pi r^2.$$

Therefore,

$$\dfrac{dA}{dr} = -\dfrac{336}{r^2} + 4\pi r = 0$$

if $r = \sqrt[3]{84/\pi} \approx 2.99$ cm. Since $(d^2 A/dr^2 > 0$ if r has this value, it corresponds to a local minimum of the function $A(r)$.) It follows that $h = 2\sqrt[3]{84/\pi} \approx 5.98$ cm.

27. Let x be the length of a side of the base and h be the height of the box. Then $100 = 8x + 4h$, so $h = 25 - 2x$.

    (a) The volume of the box is $x^2 h = x^2(25 - 2x) = V(x)$. Since $V'(x) = 50x - 6x^2$ and $V''(x) = 50 - 12x$, the volume is maximized if $x = 25/3$; $V(25/3) = 15625/27 \approx 578.7$ cm^3.

    (b) The surface area of the box is $2x^2 + 4xh = 2x^2 + 4x(25 - 2x) = 100x - 6x^2 = A(x)$. Since $A'(x) = 100 - 12x$ and $A''(x) = -12$, the surface area is maximized if $x = 25/3$; $A(25/3) = 1250/3$ cm^2.

29. (a) Both conditions are satisfied if $x \in [0, 6]$.

    (b) The results in part (a) imply that the dam can be built at most 6 miles downstream. If the dam were constructed at this point, it would be $W(6) = 100$ feet wide and $D(6) = 130$ feet high.

    (c) $W(x)$ achieves its maximum value on the interval $[0, 6]$ at $x = 0$; $W(0) = 220$.

    (d) $W(x)$ achieves its minimum value on the interval $[0, 6]$ at $x = 4$; $W(4) = 60$. [$W'(x) = 20(x - 4)$ and $W''(x) = 20$.]

    (e) The cost of building the dam x miles downstream is
    $C(x) = kW(x)D(x) = 100k \left(2x^3 - 15x^2 + 36x + 22\right)$, where k is a positive constant. Now, $C'(x) = 600k \left(x^2 - 5x + 6\right)$ so $x = 2$ and $x = 3$ are stationary points of C. Finally, since $C(0) < C(3) < C(2) < C(6)$, the dam should be built $x = 0$ miles downstream.

31. If x is the length of an edge of the square and r is the radius of the circle, then $L = 4x + 2\pi r$. We wish to maximize $S = x^2 + \pi r^2$. Since $r = (L - 4x)/2\pi$, the equation for S can be written in the form $S(x) = x^2 + (L - 4x)^2/4\pi$, where $0 \le x \le L/4$. Now, $S'(x) = 2x - 2(L - 4x)/\pi = 0$ if $x = L/(4 + \pi)$. Since $S''(x) = 2 + 8/\pi > 0$, this value of x corresponds to a local minimum of S! Thus, the maximum value of S must occur when $x = 0$ or when $x = L/4$. Since $S(0) = L^2/4\pi$ and $S(L/4) = L^2/16$, it follows that the sum of the areas is maximized when all of the wire is used to form a circle.

§4.4 Parametric Equations, Parametric Curves

1. The curve is the upper unit semi-circle plotted from $(-1, 0)$ to $(0, 1)$ to $(1, 0)$.

3. The curve is the right unit semi-circle plotted from $(0, -1)$ to $(1, 0)$ to $(0, 1)$.

5. The curve is the unit circle plotted clockwise from $(0, -1)$ to $(0, 1)$ to $(0, -1)$.

7. In each case the idea is to calculate $\sqrt{f'(t)^2 + g'(t)^2}$; if the result is constant, then the curve has constant speed. Among the given choices only the last—$x = \sin(\pi t)$, $y = \cos(\pi t)$—has constant speed.

9. (a) The spacing of bullets suggests that P moves quickly at $t = 3$, $t = 4$, $t = 9$, and $t = 10$, and slowly at $t = 0$, $t = 1$, $t = 6$, and $t = 7$.

   (b) The distance along the curve from $t = 2.5$ to $t = 3.5$ seems to be about 3 units. Thus P appears to travel about 3 units per second at $t = 3$.

   (c) Use the curve to estimate the speed of P at $t = 6$. The distance along the curve from $t = 5.5$ to $t = 6.5$ seems to be about 1 unit. Thus P appears to travel about 1 unit per second at $t = 6$.

11. (a) The result is the circle of radius 2, centered at $(2, 1)$.

    (b) Here's the calculation: Since $x = a + r\cos t$ and $y = b + r\sin t$,
    $$(x - a)^2 + (y - b)^2 = r^2(\cos t)^2 + r^2(\sin t)^2 = r^2.$$

    (c) Setting $x = 2 + \sqrt{13}\cos t$, $y = 3 + \sqrt{13}\sin t$, and $0 \leq t \leq 2\pi$, gives the circle of radius $\sqrt{13}$, centered at $(2, 3)$.

    (d) No proper "curve" results: for all t, (x, y) stays put at $(2, 3)$.

13. (a) The origin corresponds to $t = 0$; $P(0.1) \approx (0.48, 0.56)$; $P(\pi/2) = (1, 0)$. Thus P starts at the origin and starts off in a northeasterly direction.

    (b) Both x and y are 0 if and only if both $5t$ and $6t$ are integer multiples of π. This occurs only for $t = 0$, $t = \pi$, and $t = 2\pi$.

    (c) Using the t-interval $0 \leq t \leq 4\pi$ would produce exactly the same curve, but it would be traversed twice.

15. (a) The curve starts at $(at_0 + b, ct_0 + d)$ and ends at $(at_1 + b, ct_1 + d)$.

    (b) $y = \dfrac{c}{a}(x - b) + d$

    (c) $x = \dfrac{a}{c}(y - d) + b$

    (d) If $a = c = 0$, the parametric curve is just the point (b, d).

17. (a) The model would be more realistic if it took wind resistance into account. To do so, one would need some mathematical information about wind resistance.

    (b) Imitate the argument given for $f(t)$. Notice, too, that if $g(t) = 7 - 16t^2$, then $g'' = -32$, $g(0) = 7$, and $g'(0) = 0$, just as claimed.

    (c) By definition, $s(t) = \sqrt{f'(t)^2 + g'(t)^2} = \sqrt{150^2 + (-32t)^2} = \sqrt{22500 + 1024t^2}$. Plotting this function over the interval $0 \leq t \leq 0.661$ (when the ball hits the ground) gives almost a horizontal line—the velocity changes very little over the short time interval.

19. (a) If $x = f(t) = s_0 t$ and $y = g(t) = 7 - 16t^2$ it's easy to check directly that $f''(t) = 0$, $f'(0) = s_0$, $f(0) = 0$, $g''(t) = -32$, $g'(0) = 0$, and $g(0) = 7$. These are the necessary conditions.

    (b) The ball reaches home plate when $f(t) = s_0 t = 60.5$, i.e., at $t = 60.5/s_0$ seconds.

SECTION 4.4 PARAMETRIC EQUATIONS, PARAMETRIC CURVES

(c) The trajectory is parabolic for any $s_0 > 0$. (If $s_0 = 0$, the ball drops straight down.) This can be seen by eliminating t. Since $x = s_0 t$, $t = x/s_0$, so $y = 7 - 16t^2 = 7 - 16x^2/s_0^2$. This is the equation of a parabola in the xy-plane.

21. Now, $x = 200 \ln(3t/4 + 1)$.

(a) $x(t) = 60.5$ at time $t = 4\left(e^{121/400} - 1\right)/3 \approx 0.47098$. Thus, the air-dragged ball takes about 0.0677 seconds longer to reach the plate.

(b) $y(t) \approx 3.4508$ feet at the time when $x(t) = 60.5$

(c) When $x = 60.5$, the ball's speed is approximately 111.87 ft/sec.

(d) When $y = 0$, $x \approx 80.569$ feet.

§4.5 Related Rates

1. (a) $x + 2y = 3 \implies x'(t) + 2y'(t) = 0 \implies x'(t) = -2y'(t)$.
   (b) Using part (a), $1 = -2y'$ so $y' = -1/2$.
   (c) The line $x + 2y = 3$ has slope $-1/2$. Thus, a change of Δx in x leads to a change $\Delta y = -\Delta x/2$. In other words, the rate of change of y is $-1/2$ the rate of change of x for all x.

3. (a) Using implicit differentiation,
   $x^2(t) + y^2(t) = 1 \implies 2x(t)x'(t) + 2y(t)y'(t) = 0 \implies x(t)x'(t) + y(t)y'(t) = 0$ for all t.
   (b) If $x(0) = 1$ and $y'(0) = 1$, then $y(0) = 0$ and $x'(0) = 0$. This means that the moving point is at $(1, 0)$ at this time. Since $x'(0) = 0$ and $y'(0) = 1$, the point is moving vertically upwards.
   (c) If $x(t_0) = 1/2$, the equation $x^2 + y^2 = 1$ implies that $y(t_0) = \pm\sqrt{3}/2$. If $x(t_0) = 1/2$ and $x'(t_0) = 1$, the equation in part (a) implies that $1/2 \pm \dfrac{\sqrt{3}}{2} y'(t_0) = 0$. Thus, if $y(t_0) = \sqrt{3}/2$, $y'(t_0) = -\sqrt{3}/3$; if $y(t_0) = -\sqrt{3}/2$, $y'(t_0) = \sqrt{3}/3$.
   (d) At the time t_0, the moving point is at $(1/2, \sqrt{3}/2)$ or at $(1/2, -\sqrt{3}/2)$. If it is at the first point, then it is moving to the right and downward. If it is at the second point, then it is moving to the right and upward.

5. Using similar triangles,
$$\frac{12}{x+s} = \frac{6}{s} \implies 12s = 6x + 6s \implies 6s = 6x \implies s = x$$
for every time t. Thus, if Hal is 30 feet from the lamppost (i.e., $x = 30$), the length of his shadow is $s = 30$ feet. Furthermore, since $s'(t) = x'(t)$ and $x'(t) = 7$ feet per second, his shadow is lengthening at the rate of 7 feet per second.

7. If $L(t)$ and $W(t)$ are the length and width of the rectangle at time t, then the area of the rectangle at time t is $A(t) = L(t) \cdot W(t)$ and $A'(t) = L'(t)W(t) + L(t)W'(t)$ is the rate of change of the rectangle's area. Plugging in the values given in the problem, we find that the area of the rectangle is increasing at a rate of 22 cm^2/sec.

9. Let $E(t)$ be the distance from the intersection of the bicyclist traveling west and $S(t)$ be the distance from the intersection of the bicyclist traveling south. From the information given in the problem, we have $E = 4$ miles, $E' = -9$ miles/hour, $S = 3$ miles, and $S' = 10$ miles per hour. The distance between the two bicyclists $D(t)$ at any time can be determined from the equation $D^2 = E^2 + S^2$. Differentiating both sides of this equation with respect to time (using the product rule), we find that
$$2D \cdot D' = 2E \cdot E' + 2S \cdot S'$$
Now, at the time of interest $D = 5$ miles, so we may use the previously given values of E, E', S, and S', so $5 \cdot D' = 4 \cdot (-9) + 3 \cdot 10$. This implies that $D' = -\frac{6}{5}$ miles per hour. Therefore, the distance between the bicyclists is *decreasing* at a rate of 1.2 miles per hour.

11. Let $x(t)$ be the distance from the runner to first base at time t. Then the distance from the runner to second base is $D(t) = \sqrt{90^2 + x(t)^2}$ and $D'(t) = x(t)x'(t)/D(t)$. When the runner is halfway to first base $x(t) = 45$ and $x'(t) = -20$ ft/sec, so $D'(t) = -20/\sqrt{5} = -4\sqrt{5}$ ft/sec.

13. The area of the ring between the two circles is increasing.

    Let $r(t)$ be the radius of the inner circle at time t and $R(t)$ be the radius of the outer circle at time t. Then, the area of the ring between the two circles is $A(t) = \pi\left((R(t))^2 - (r(t))^2\right)$. Therefore, $A'(t) = 2\pi(R(t)R'(t) - r(t)r'(t))$. At the time when $R = 10$, $R' = 2$, $r = 3$, and $r' = 5$, $A' = 10\pi$. Since this value (the rate of change of the area of the ring between the two circles) is positive, the area is increasing.

SECTION 4.5 RELATED RATES

15. Since both flights are at the same elevation, we may describe the positions of the planes in terms of just their x- and y-coordinates. The coordinates of the Pachyderm plane t hours after observation are $(0, -36 + 410t)$ and the coordinates of the Peterpan plane are $(41 - 455t, 0)$. Thus, the distance between the two planes at time t is $D(t) = \sqrt{(-36 + 410t)^2 + (41 - 455t)^2}$.

   (a) At the time of closest approach the planes are $\sqrt{7396/15005} \approx 0.702$ nautical miles apart.

   (b) At time $t = 6683/75025 \approx 0.0891$ hours, $D'(t) = 0$ and $D''(t) > 0$. Thus, the controllers have approximately 5.345 minutes before the time of closest approach.

17. Since the slick has the form of a circular cylinder, the volume of oil in the slick is $V = Ah$ where A is the area of the slick and h is its depth. At the moment of time described in the problem, the slick has area $A = \pi r^2 = \pi 500^2$ square feet, $h = 0.01$ feet, and $h' = -0.001$ feet/hour. Since $V' = A'h + Ah' = -5$ cubic feet per hour, the surface area is increasing at a rate of $A' = (-5 + 250\pi)/0.01 \approx 78{,}040$ square feet per hour.

19. (a) The volume of the ice is $V = 4\pi \left(3R^2T + 3RT^2 + T^3\right)/3$ where $R = 4$ inches is the radius of the iron ball and $T = 2$ inches is the thickness of the ice. Now, the rate of change of the volume of the ice can be related to the rate of change in the thickness of the ice by differentiating: $V' = 4\pi \left(2RR'T + R^2T' + 2RTT' + R'T^2 + T^2T'\right)$. Since $R' = 0$ and $V' = -10$ in^3/min, we may solve for $T' = -10/144\pi$ in/min.

   (b) The surface area of the ice is $S = 4\pi (R + T)^2$. Thus, the rate of change of the surface area is $S' = 8\pi (R + T)(R' + T') = 8\pi \cdot 6 \cdot T' = -10/3$ in^2/min.

21. (a) Home plate, the position of the ball, and first base can be considered to be the vertices of a right triangle. Let S be the length of the hypotenuse of this right triangle and F and T be the lengths of the other two sides. At the instant of time when the ball is halfway to third base $F = 90$ feet and $T = 45$ feet. Since $S^2 = F^2 + T^2$, $S = \sqrt{10125}$ feet. Differentiating both sides of the equation relating S to F and T, we obtain $2SS' = 2FF' + 2TT'$, or $SS' = FF' + TT'$. Now, $T' = 100$ ft/sec and $F' = 0$, so $S' = (45 \cdot 100)/\sqrt{10125} \approx 44.721$ ft/sec.

   (b) This is similar to part (a) except that F is now the distance between home plate and the runner. The ball reaches the point halfway to third base in 0.45 seconds, so $F = 25$ ft/sec $\cdot$ 0.45 sec $= 11.25$ feet. Thus, $S' = (11.25 \cdot 25 + 45 \cdot 100)/\sqrt{11.25^2 + 45^2} \approx 103.08$ ft/sec.

23. The volume of a cone with radius r and height h is $V = \pi r^2 h/3$. If R is the ratio h/r, then $V = \pi h^3/3R^2$. Therefore, $V'(t) = \pi h(t)^2 h'(t)/R^2$. Since $V'(t) = -10$ cm^3/min when $h'(t) = -2$ cm/min and $h(t) = 8$ cm, we find that $R^2 = 8^2\pi/5$ so $R = 8\sqrt{\pi/5}$.

25. If t is measured in hours and $t = 0$ corresponds to 12:00, the coordinates of the tip of the minute hand are $(x_m(t), y_m(t))$ where $x_m(t) = 7\cos(2\pi t - \pi/2) = 7\sin(2\pi t)$ and $y_m(t) = 7\cos(2\pi t)$. Similarly, the coordinates of the tip of the hour hand are $(x_h(t), y_h(t))$ where $x_h(t) = 5\cos(\pi t/6)$ and $y_h(t) = 5\sin(\pi t/6)$. Thus, the distance between the tips of the hands at time t is

$$D(t) = \sqrt{\left(x_m(t) - x_h(t)\right)^2 + \left(y_m(t) - y_h(t)\right)^2}$$
$$= \sqrt{74 - 70\left(\sin(2\pi t)\cos(\pi t/6) + \cos(2\pi t)\sin(\pi t/6)\right)}$$
$$= \sqrt{74 - 70\sin(13\pi t/6)}$$

so the distance between the hands at time t is changing at the rate

$$D'(t) = \frac{770\pi}{12 \cdot D(t)} \sin\left(\frac{13\pi}{6}t\right)$$

Thus, at time $t = 9$ the distance between the hands is increasing at the rate of
$D'(9) = 770\pi/12\sqrt{74} \approx 23.434$ feet/hour or approximately 4.6868 inches/minute or 0.3906 feet/minute.

27. The elevation of the rocket at time t is $y = 100 \tan \theta$ where θ is the angle of elevation at time t. Therefore, since both y and θ are functions of time, the speed of the rocket at time t is $y'(t) = 100 \cdot \theta'(t) \cdot \sec^2 \theta(t)$ where θ' is rate of change in the angle of elevation at time t. From the problem statement, we are interested in the value of y' at the time when $\theta = \pi/3$ radians and $\theta' = \pi/15$ radians/sec. Thus, at this time, $y' = 100 \cdot 4 \cdot \pi/15 = 80\pi/3 \approx 83.776$ m/sec.

29. When the water in the tank is h feet deep, the volume of water in the tank is
$V = \frac{1}{2}(3 + (3 + h))h \cdot 10 = 30h + 5h^2$ cubic feet. (The volume is the cross-sectional area times the length of the tank. When the water has height h, the upper base of the trapezoidal cross-section has length $3 + h$.) Thus, $V' = 10(3 + h)h'$ cubic feet/minute. When $h = 1$ foot and $h' = 1/48$ feet/minute, $V' = 5/6$ cubic feet/minute.

SECTION 4.6 NEWTON'S METHOD: FINDING ROOTS

§4.6 Newton's Method: Finding Roots

1. The first three answers, written as fractions, are $x_1 = 9/4$, $x_2 = 161/72$, and $x_3 = 51841/23184$. (The last answer, by the way, is correct to 9 decimal places!)

3. Newton's method will converge to the leftmost root (i.e., the root near -1.88) because the x-intercept of the line tangent to the graph at $x = 0.95$ is to the left of the local maximum point at $x = -1$. (In fact, the x-intercept of this tangent line is approximately -2.44.) It follows that subsequent iterations of Newton's method will converge to the root near -1.88 (all tangent lines based at points to the left of $x = -1$ have x-intercepts to the left of $x = -1$).

5. (a) Newton's method with $x_0 = 0.5$ gives $x_1 = 0.724638$, $x_2 = 0.7063515$, and $x_3 = 0.706115$. Thus, to three-decimal-place accuracy, the solution is $x = 0.706$.

   (b) The function f has no other roots, because $f'(x) = 5x^4 + 4 > 0$ for all x. (This means that f is increasing everywhere.)

7. (a) $N(x) = x - \dfrac{x^2 - a}{2x} = \dfrac{x}{2} + \dfrac{a}{2x}$.

   (b) If $x > \sqrt{a}$, then $a/x < a/\sqrt{a} = \sqrt{a}$. If $x < \sqrt{a}$, then $a/x > a/\sqrt{a} = \sqrt{a}$.

   (c) The result follows from simple algebra.

   (d) The estimates are $1, 3/2, 17/12, 577/408, 665857/470832$.

9. (a) The first few Newton estimates are $1.250000000, 1.0250000001.000304878, 1.000000046$. They are accurate to 0, 1, 3, and 7 decimal places respectively.

   (b) Newton's method finds $x = 1$ if $x_0 > 0$; it finds $x = -1$ if $x_0 < 0$. It fails if $x_0 = 0$.

11. x_{n+1} is the x-intercept of the line tangent to the graph of f that passes through $(x_n, f(x_n))$; this line has slope $f'(x_n)$. Therefore, equation ?? implies that $x_{n+1} = x_n - f(x_n)/f'(x_n)$.

13. (a) Suppose that $0 < x < \sqrt{a}$. Then $\sqrt{a}x < \sqrt{a}\sqrt{a} = a \implies \sqrt{a} < a/x$. Alternatively, suppose that $x > \sqrt{a} > 0$. Then $x > \sqrt{a} \implies \sqrt{a}x > a \implies \sqrt{a} > a/x$.

    (b) $N(x) = x - \dfrac{x^2 - a}{2x} = x - x/2 + a/2x = (x + a/x)/2$.

    (c) If $x = \sqrt{a}$, then $N(x) = x$. In other words, Newton's method "stops"—as it should—when it finds the *exact* root $\sqrt{a}$.

15. (a) The approximate roots are $-0.244817, 3.80675$, and 6.43807.

    (b) Newton's method jumps back and forth between the estimates $x = 2$ and $x = 5$. Since $f(2)/f'(2) = -3$, and $f(5)/f'(5) = 3$, applying Newton's method to 2 gives 5; applying it to 5 gives 2. (Draw the graph to see the situation more clearly.)

    (c) $f'(1.39)$ is a small number because $x = 1.39$ is near a critical point of f (i.e., the tangent line is nearly horizontal). This means that the x-intercept of the tangent line may be far from the current estimate of the root. This causes Newton's method to converge slowly.

17. To find the minimum value of $g(x)$, consider $g'(x) = -20x^{-3} + 6x + 1$. For $x > 0$, $g''(x) = 60x^{-4} + 6 > 0$, so g is concave up for x in $[1, 10]$. Thus g has at most one local minimum on $[1, 10]$; it must occur at the one place where where $g'(x) = 0$. (Since $g''(x) > 0$, $g'(x)$ is always increasing, so $g'(x)$ can equal 0 for at most one value of $x > 0$.)

    Applying Newton's method to $g'(x)$, starting from $x_0 = 2$, locates the root $x \approx 1.3114$. Thus $g(1.3114) \approx 8.285$ is the minimum value of g.

19. From a graph, it appears that f achieves its maximum value between $x = 2.5$ and $x = 3$. To find the maximum value of f, therefore, we need to identify the corresponding root of f'. To do this, we use the Newton iteration formula $x_{n+1} = x_n - \dfrac{f'(x_n)}{f''(x_n)}$, where $f'(x) = 2x\sin(x^2) + 2x^3\cos(x^2)$ and $f''(x) = 2\sin(x^2) + 4x^2\cos(x^2) + 6x^2\cos(x^2) - 4x^4\sin(x^4)$. Using $x_0 = 3$, we obtain $x_1 = 2.78236$, $x_2 = 2.82791$, $x_3 = 2.82467$, $x_4 = 2.82465$, and $x_5 = 2.82465$. Thus, the maximum value of f is $f(2.82465) = 7.91673$.

21. The Newton's estimates "blow up." (Draw the graph to see why.) The underlying reason is that if $f(x) = x^{1/3}$, then (as algebra shows), $x - f(x)/f'(x) = -2x$.

§4.7 Building Polynomials to Order; Taylor Polynomials

1. The value and derivatives are (in order), 1, 1, 2, 6, 24, 120, 0, 0. Note that all derivatives beyond the fifth are zero.

3. (a) $P_2(x) = 1 + 2(x - 1) + (x - 1)^2$.

   (b) Multiplying out $P_2(x)$ gives x^2. This happens because the quadratic approximation to a quadratic function f is f itself.

5. $f'(x) = \frac{1}{3}x^{-2/3}$ and $f''(x) = -\frac{2}{9}x^{-5/3}$. Therefore, since, $f(8) = 2$, $f'(8) = 1/12$, and $f''(8) = -1/144$,

   $$P_2(x) = 2 + \frac{1}{12}(x - 8) - \frac{1}{288}(x - 8)^2 = \frac{10}{9} + \frac{5}{36}x + \frac{1}{288}x^2.$$

7. Theorem ?? says $p_2(x) = 2 + b_2(x - 1)^2 + b_3(x - 1)^3$. The conditions $p_2(2) = 1$ and $p'_2(2) = 0$ imply together that $b_2 = -3$ and $b_3 = 2$.

9. If $f(x) = \dfrac{1}{1-x}$, $n = 3$, and $x_0 = 0$, then $P_3(x) = 1 + x + x^2 + x^3$.

11. If $f(x) = \ln x$, $n = 3$, $x_0 = 1$, then $P_3(x) = (x - 1) - \dfrac{(x-1)^2}{2} + \dfrac{(x-1)^3}{3}$.

13. If $f(x) = \sqrt{x}$, $n = 3$, $x_0 = 4$, then $P_3(x) = 2 + \dfrac{x-4}{4} - \dfrac{(x-4)^2}{64} + \dfrac{(x-4)^3}{512}$.

15. $\ell(x) = 1$; $q(x) = 1 - x^2/2$

17. $f(x) = e^x$ has $\ell(x) = 1 + x$, $q(x) = 1 + x + x^2/2$.

19. $f(x) = \arcsin x$ has $\ell(x) = x$, $q(x) = x$.

21. (a) The graph of an odd function is symmetric about the origin. All graphs shown have that property.

    (b) The *even-order* Maclaurin polynomials P_2, P_4, P_6, and P_8 are the same as the odd-order Maclaurin polynomials P_1, P_3, P_5, and P_7. (This happens because the sine function is odd. So, therefore, are all of its Maclaurin polynomials.)

23. (a) We'll use the linear approximation $l(x)$ at $x = 1$. Since $f(1) = 0$ and $f'(1) = \sin 1 \approx 0.84147$, $l(x) = 0 + \sin 1(x - 1)$; $l(0.5) = 0 + \sin 1(-0.5) \approx -0.42074$.

    (b) Whether the estimate above is too big or too small depends on the concavity of f between $x = 0.5$ and $x = 1$. Notice that $f''(x) = 2x \cos(x^2)$; thus $f''(x) > 0$ for x in $[-0.5, 1]$; so f is concave *up*, and so the linear approximation *underestimates* f.

    (c) $f''(1) = 2\cos 1 \approx 1.08060$; therefore the quadratic approximation at $x = 1$ has the form $q(x) = f(1) + f'(1)(x - 1) + \frac{f''(1)}{2}(x - 1)^2 = 0 + \sin 1(x - 1) + \cos 1(x - 1)^2$. Therefore $q(0.5) \approx -0.28566$.

25. Let $f(x) = \sqrt{x}$ and $x_0 = 100$. Then $f'(x) = 1/2\sqrt{x}$, $f(x_0) = 10$, and $f'(x_0) = 1/20$. Therefore, $f(103) = f(x_0 + 3) \approx f(x_0) + 3f'(x_0) = 10 + 3/20 = 10.15$. A calculator gives $\sqrt{103} \approx 10.14889$; the difference is $10.15 - 10.14889 = 0.00111$.

27. Let $f(x) = \sin x$ and $x_0 = \pi/3$. Then $f'(x) = \cos x$, $f(x_0) = \sqrt{3}/2$ and $f'(x_0) = 1/2$. Therefore, $\sin 58° = \sin(\pi/3 - \pi/90) \approx f(x_0) - (\pi/90)f'(x_0) = \sqrt{3}/2 - \pi/180 \approx 0.84857$. A calculator gives $\sin 58° \approx 0.84805$; the difference is $0.84857 - 0.84805 = 0.00052$.

29. (a) $\ell_p(t) = 25t$; $\ell_p(1) = 25$; $\ell_p(-1) = -25$

    (b) $q_p(t) = 25t + t^2$; $q_p(1) = 26$; $q_p(-1) = -24$

(c) $\ell_v(t) = 25 + 2t$; $\ell_v(1) = 27$

31. (a) $\ell(t) = 100$ meters; $\ell(1) = 100$ meters
    (b) $q(t) = 100 - 4.9t^2$ meters; $q(1) = 95.1$ meters

33. (a) $f(1) = 1$, $f'(1) = 0$
    (b) no, $q(1) = 3 \neq 1$
    (c) no, $r'(1) = -4 \neq 0$
    (d) no, $s''(1) = 4 > 0$ but $f''(1) < 0$

35. No. Since f has a local maximum at $x = 1$, $f'(1) = 0$. Since $p'(1) \neq 0$, p cannot be a Taylor polynomial for f.

37. The polynomial $p(x) = x - x^3/6$ is the fourth-order Taylor polynomial approximation to $f(x) = \sin x$ based at $x_0 = 0$. Therefore, $p(0) = f(0)$, $p'(0) = f'(0)$, $p''(0) = f''(0)$, $p'''(0) = f'''(0)$, and $p^{(4)}(0) = f^{(4)}(0)$.

39. Let $f(x) = \sin x$ and $g(x) = x(\cos x)^{1/3}$. The graphs of f and g are almost indistinguishable in a viewing window centered at the origin because $f(0) = g(0)$, $f'(0) = g'(0)$, $f''(0) = g''(0)$, $f'''(0) = g'''(0)$, and $f^{(4)}(0) = g^{(4)}(0)$. [NOTE: $f^{(5)}(0) \neq g^{(5)}(0)$.]

41. Let $f(x) = e^x$ and $g(x) = 5/2 - \frac{5}{3}\cos x + \frac{1}{6}\cos(2x) + \frac{5}{3}\sin x - \frac{1}{3}\sin(2x)$. The graphs are almost indistinguishable in a viewing window centered at the point $(0, 1)$ because $f(0) = g(0)$, $f'(0) = g'(0)$, $f''(0) = g''(0)$, $f'''(0) = g'''(0)$, and $f^{(4)}(0) = g^{(4)}(0)$. [NOTE: $f^{(5)}(0) \neq g^{(5)}(0)$.]

43. (a) $q_{100}(x) = 1$
    (b) $q_{010}(x) = x$
    (c) $q_{001}(x) = x^2/2$
    (d) The graphs are two lines and a parabola.
    (e) If $q = aq_{100}(x) + bq_{010}(x) + cq_{001}(x)$, then
    $q(0) = aq_{100}(0) + bq_{010}(0) + cq_{001}(0) = a \cdot 1 + 0 + 0 = a$; the other parts are similar.

45. (a) Yes, they are inflection points, since $p_1''(1/2) = 0$ and $p_2''(3/2) = 0$.
    (b) Theorem ?? says that $p_3(x) = 2 + 0(x - 1) - 6(x - 1)^2 + a_3(x - 1)^3$. To find a_3 we use $p_3(2) = 1$. But $p_3(2) = 2 - 6 + a_3 = 1 \implies a_3 = 5$, so $p_3(x) = 2 - 6(x - 1)^2 + 5(x - 1)^3$.
    (c) The inflection point on p_3 is at $x = 7/5$.

SECTION 4.8 WHY CONTINUITY MATTERS

§4.8 Why Continuity Matters

1. The graph of f has points of discontinuity at $-3, -1, 1$, and 2. Thus, f is continuous on the intervals $[-4, -3), (-3, -1), (-1, 1), (1, 2)$, and $(2, 4)$.

3. The function $h(x) = f(x) + g(x)$ will be continuous wherever *both* f and g are continuous. Thus, points of discontinuity of either f or g are possible points of discontinuity of h. By examining values of h near the points $-3, -1, 1, 2$, and 3, we find that h is continuous on the intervals $[-4, -3), (-3, -1), (-1, 1), (1, 2), (2, 3)$, and $(3, 4)$.

5. The function $h(x) = f(x)/g(x)$ will be continuous wherever $g(x) \neq 0$ and both f and g are continuous. Thus, the points $-3, -1, -1/3, 1, 2$, and 3 are possible points of discontinuity. By examining values of h near these points, we find that h is continuous on the intervals $(-4, -3), (-3, -1), (-1, -1/3), (-1/3, 1), (1, 2), (2, 3)$, and $(3, 4)$.

7. Yes, there is an input to f that produces every output value between $f(-4) = -2$ and $f(4) = 1$.

9. Yes, $-2 = f(-4) \leq f(x) \leq f(0) = 2$ if $-4 \leq x \leq 4$. That is, f assumes both a maximum value and a minimum value over the interval $[-4, 4]$.

11. Since f is continuous on the interval $[0.1, 1]$, the EVT says that f assumes both a minimum value and a maximum value over this interval.

13. The interval $(0, 1]$ is not a closed interval. Thus, since the hypotheses of the EVT are not satisfied, f need not assume a maximumm and minimum value on this interval.

15. The function f is not continuous on the interval $[-1, 1]$, so the hypotheses of the theorem are not satisfied.

17. Since f is a polynomial, it is continuous everywhere. Moreover, $f(0) = 2$ and $f(1) = -1$. Therefore, since $-1 < 0 < 2$, the IVT implies that f has a root somewhere in the interval $(0, 1)$.

19. Note that $1^2 < 3 < 2^2$, so $\sqrt{3}$ lies in the interval $[1, 2]$; the midpoint of this interval is $m_1 = 3/2$. Since $(m_1)^2 = 9/4 < 3$, $\sqrt{3}$ lies in the interval $[3/2, 2]$; the midpoint of this interval is $m_2 = 7/4$. Since $(m_2)^2 = 49/16 > 3$, $\sqrt{3}$ lies in the interval $[3/2, 7/4]$; the midpoint of this interval is $m_3 = 13/8$. Since $(m_3)^2 = 169/64 < 3$, $\sqrt{3}$ lies in the interval $[13/8, 7/4]$; the midpoint of this interval is $m_4 = 27/16$.

21. (a) Each iteration of the bisection method reduces the width of the interval in which the root is guaranteed to be located by a factor of 2. Thus, after 1 iteration, the interval has length $1/2^1 = 1/2$; after 2 iterations, the interval has length $1/2^2 = 1/4$; and so forth. The smallest integer n for which $1/2^n < 10^{-2}$ is $n = 7$. Therefore, 7 iterations of the bisection method are necessary to guarantee an estimate of the root within 10^{-2}.

    (b) The smallest integer n for which $1/2^n < 10^{-3}$ is $n = 10$. Therefore, 10 iterations of the bisection method are necessary to guarantee an estimate of the root within 10^{-3}.

    (c) The smallest integer n for which $1/2^n < 10^{-5}$ is $n = 17$. Therefore, 17 iterations of the bisection method are necessary to guarantee an estimate of the root within 10^{-5}.

    (d) The smallest integer n for which $1/2^n < 10^{-10}$ is $n = 34$. Therefore, 34 iterations of the bisection method are necessary to guarantee an estimate of the root within 10^{-10}.

23. For each distance between 37 and 12 miles, there was some time at which I was that far from home. (Distance is a continuous function of time.)

25. The IVT says nothing—the amount of money in my pocket is not a continuous function of time, since only 2-decimal place numbers are possible.

27. Let $g(x) = f(x) - x$. Then, g is a continuous function, $g(0) > 0$, and $g(1) < 0$. The IVT implies that g must have a root in the interval $[0, 1]$, so $f(x) - x = 0 \implies f(x) = x$ for some x in the interval $[0, 1]$.

29. Since $f(0) \cdot f(1) < 0$, the two factors must have opposite signs. Therefore, the IVT guarantees that $f(x) = 0$ for some x in the interval $(0, 1)$.

31. The statement **cannot** be true since $f(x) \leq 5$ for all x.

33. The statement **must** be true because, by hypothesis, there is a value of x for which $f(x) = -7 \implies |f(x)| = 7$.

35. The statement **might** be true. If f is continuous on $[1, 5]$, then the IVT guarantees that the statement is true. However, if f isn't continuous, the statement might be true, but it need not be.

37. The statement **might** be true.

    Notice first that
    $$f(1) \cdot f(5) < 0 \iff f(1) \text{ and } f(5) \text{ have opposite signs.}$$

    Thus the question is whether having a root in $(1, 5)$ means, necessarily, that f changes sign from $x = 1$ to $x = 5$. A little thought shows that f may or may not do so. E.g., $f(x) = x - 3$ *does* change sign from $x = 1$ to $x = 5$, but $f(x) = (x - 3)^2$ *doesn't*.

39. Since f is a polynomial, it is continuous everywhere. Also, since $f(0) = 2$ and $f(-1) = -5$, the IVT guarantees that a root of f lies in the interval $(-1, 0)$.

41. Consider the function $g(x) = f(x) - x$. Finding a *fixed point* of f is the same thing as finding a *root* of g, since $g(x) = 0 \iff f(x) = x$. So let's show that g has a root.

    By hypothesis, $g(0) = f(0) \geq 0$, and $g(1) = f(1) - 1 \leq 1 - 1 \leq 0$. Thus $g(0) \geq 0 \geq g(1)$.

    If either $g(0) = 0$ or $g(1) = 0$ we've *found* our root, and we're done. The only alternative is that $g(0) > 0 > g(1)$. In this case, the IVT guarantees that g has a root somewhere in $(0, 1)$, so we're done.

§4.9 Why Differentiability Matters; The Mean Value Theorem

1. The converse of the given statement is "If it's cloudy, then it's raining." The given statement is true, but its converse is false.

3. The converse of the given statement is "If $x > 1$, then $x > 2$." The given statement is true, but its converse is false.

5. The converse of the given statement is "If $f(x) = x^2$, then $f'(x) = 2x$." The given statement is false, but its converse is true.

7. If $m = (a+b)/2$, then $\big(q(b) - q(a)\big)/(b - a) = q(m)$.

9. No such function exists. Suppose that x_1, x_2, and x_3 are roots of f. Rolle's theorem implies that f must have a stationary point between x_1 and x_2 and also between x_2 and x_3 since $f(x_1) = f(x_2) = f(x_3) = 0$. Thus, f must have at least 2 stationary points.

11. The sine function is an example of a function with infinitely many roots and infinitely many stationary points.

13. Suppose that $f'(x) > 2$ for all x. Then the speed limit law implies that $f(1) - f(0) > 2$. This contradicts the fact that $f(1) - f(0) = 1$. Therefore, it is impossible to find a differentiable function f with the desired properties.

15. No, f is not differentiable at $x = 0$.

17. Yes, f is continuous on the closed interval $[1, 2]$ and differentiable on the open interval $(1, 2)$.

19. Take a look at the graph of $f'(x)$ on the interval $[-1, 1]$. You will see that it has a value of ≈ 8 at $x = -1$, and a value of 0 at $x = 1$. Since $f'(x)$ is a polynomial and therefore continuous, by the IVT there must exist a c on $[-1, 1]$ where $f'(c) = 2$. Since $f'(x)$ must equal 2 at some c, then $f(x)$ must have slope 2 at that same c.

21. The MVT says that $f'(c) = \big(f(2) - f(1)\big)/(2 - 1) = 3$ for some c in the interval $(1, 2)$. Since $f'(x) = 2x$, the only suitable value of c is $c = 1.5$.

23. Since g is not differentiable at 0, Rolle's theorem doesn't apply.

25. If f' is a continous function, f is differentiable on $(1, 4)$. Since $f(1) = f(4) = 0$, Rolle's theorem implies that there must be a c between 1 and 4 for which $f'(c) = 0$.

27. If f' is not continuous, f' can "jump" from a positive to a negative value or *vice versa*.

29. (a) Suppose that $f(2) = 3$. Then, according to the MVT, there would be a number c such that $0 \le c \le 2$ and $f'(c) = \big(f(2) - f(0)\big)/(2 - 0) = 3/2$. However, this contradicts the hypothesis that $f'(x) < 1$. Thus, $f(2) = 3$ is not possible.

    (b) $3/2 \le f(3) \le 3$

    (c) $-1 \le f(-1) \le -1/2$

31. (a) $f'(1) = 3$, and $f'(3) = -1$.

    (b) $f(1) = 2$. The equation of the line tangent to the graph of f at $x = 3$ is $y = 5 - x$. Thus, $f(3) = 2$.

    (c) Since the function is differentiable everywhere, $f'(1) > 0$, and $f'(3) < 0$, there must, by Rolle's Theorem, be some point where $f'(x) = 0$.

    (d) To get more than one maxima on that interval, the second derivative would have to have a sign change. Since it doesn't, that means $f'(x)$ must decrease all the time, and that means it can only cross the y axis once, which in turn means that $f(x)$ can have only one critical point, which must be a maximum, due to the negative second derivative.

33. (a) Yes. The trucker traveled 100 miles in 1.25 hours. The MVT asserts that the trucker's speed must have been 80 mph at some time during the trip.

(b) The trucker's fine will be at least $125 = \$50 + \$5 \cdot 15$.

SECTION 5.1 AREAS AND INTEGRALS

§5.1 Areas and Integrals

1. The area of the entire rectangle shown is 250, so $\int_1^2 g(x)\,dx < 250$. Furthermore, since $g(x) > 0$ for all x, $\int_1^2 g(x)\,dx > 0$. Finally, it is clear from the picture that $\int_{1.75}^2 g(x)\,dx > 12.5$ (the area of one dotted rectangle). Thus, the only possible choice is $\int_1^2 g(x)\,dx \approx 45$.

3. Since $3 \leq f(x) \leq 5$ if $3 \leq x \leq 7$, $A = 12 = 3 \cdot (7-3) \leq \int_3^7 f(x)\,dx \leq 5 \cdot (7-3) = 20 = B$.
   [NOTE: Other values of A and B are possible.]

5. $\int_0^2 f(x)\,dx = 4$. This definite integral is equal to the area of a trapezoid with base 2 and heights $f(0) = 1$ and $f(2) = 3$.

7. $\int_0^5 f(x)\,dx = \int_0^2 f(x)\,dx + \int_2^5 f(x)\,dx = 4 + 9\pi/4$. (The value of the first integral on the right is the area of a trapezoid and the value of the second is the area of a one-quarter circle of radius 3.)

9. $\int_4^4 f(x)\,dx = 0$ because the left- and right-endpoints of the interval of integration are equal.

11. $\int_0^{15} f(x)\,dx = \int_0^2 |f(x)|\,dx + \int_2^5 |f(x)|\,dx + \int_5^9 |f(x)|\,dx + \int_9^{15} |f(x)|\,dx$

    $= \int_0^2 f(x)\,dx + \int_2^5 f(x)\,dx - \int_5^9 f(x)\,dx - \int_9^{15} f(x)\,dx$

    $= 4 + 9\pi/4 + 4\pi + 12 \approx 35.6$

    since $|f(x)| = f(x)$ if $f(x) \geq 0$ and $|f(x)| = -f(x)$ if $f(x) \leq 0$.

13. $\int_{12}^{15} f(x)\,dx = -3$ since the integral is the signed area of a right triangle with base 3 and altitude 2 that lies below the x-axis.

15. The value of the integral is the signed area of a rectangle with base 2 and height 3 that lies above the x-axis. Thus, $\int_{-4}^{-2} 3\,dt = 6$.

17. The value of the integral is the signed area of a trapezoid with base $5 - 2 = 3$ and heights $h_1 = 2$ and $h_2 = 5$ that lies above the x-axis. Thus, $\int_2^5 u\,du = \frac{1}{2} \cdot 3 \cdot (2+5) = \frac{21}{2}$.

19. $\int_{-3}^3 |x|\,dx = \int_{-3}^0 (-x)\,dx + \int_0^3 x\,dx$. Each of the integrals on the right side is the area of a right triangle with base 3 and altitude 3 that lies above the x-axis, so $\int_{-3}^3 |x|\,dx = 9$.

21. The value of the integral is the area of a semi-circle of radius 2 that lies above the x-axis. Thus, $\int_{-2}^2 \sqrt{4 - s^2}\,ds = 2\pi$.

23. $\int_\pi^0 r^2\,dr = -\int_0^\pi r^2\,dr = -\pi^3/3$

25. $\int_0^\pi (\pi^2 - 3t^2)\,dt = \int_0^\pi \pi^2\,dt - \int_0^\pi 3t^2\,dt = \pi^2 \int_0^\pi dt - 3 \int_0^\pi t^2\,dt = \pi^3 - 3(\pi^3/3) = 0$

27. The graph of $g(v) = (v - \pi)^2$ over the interval $\pi \leq v \leq 2\pi$ is exactly the same as the graph of $f(u) = u^2$ over the interval $0 \leq u \leq \pi$. Therefore, $\int_\pi^{2\pi} (v - \pi)^2\,dv = \int_0^\pi u^2\,du = \pi^3/3$.

29. $3 \leq 4w - w^2 \leq 4$ if $1 \leq w \leq 3$. Therefore, $A = 6 = 3 \cdot (3-1) \leq \int_1^3 (4w - w^2)\, dw \leq 4 \cdot (3-1) = 8 = B$.

31. $1/10 \leq (1+t^2)^{-1} \leq 1/2$ if $1 \leq t \leq 3$. Therefore,
$$A = 1/5 = (1/10) \cdot (3-1) \leq \int_1^3 (1+t^2)^{-1}\, dt \leq (1/2) \cdot (3-1) = 1 = B.$$

33. $\int_{-2}^{5} (3f(x) + 4g(x))\, dx = 3\int_{-2}^{5} f(x) + 4\int_{-2}^{5} g(x)\, dx = 3 \cdot 18 + 4 \cdot 5 = 74$

35. $\int_{5}^{-2} f(x)\, dx = -\int_{-2}^{5} f(x)\, dx = -18$

37. Jack's answer is too big — the area of the entire rectangle shown is only $\pi/2$. Since $\cos^8 x \geq 0$ on the entire interval $[0, \pi/2]$, the value of the integral must be positive. This rules out Ed's answer. Furthermore, the value of the integral is approximately the area of the triangle with vertices at $(0,0)$, $(0,1)$, and $(1,0)$. Since this triangle has area $1/2$, Lesley's answer is too big. Therefore, Joan's answer must be the correct answer.

    [NOTE: $\cos^8 x \leq 1$ if $0 \leq x \leq 1/2$, $\cos^8 x \leq 2/5$ if $1/2 \leq x \leq 1$, and $\cos^8 x \leq 75/10{,}000$ if $1 \leq x \leq \pi/2$. Therefore,
    $\int_0^{\pi/2} \cos^8 x\, dx = \int_0^{1/2} \cos^8 x\, dx + \int_{1/2}^{1} \cos^8 x\, dx + \int_1^{\pi/2} \cos^8 x\, dx \leq 1/2 + 1/5 + 75\pi/10{,}000 \leq 3/4.$]

39. The regions whose areas are represented by the two integrals are horizontal translates of each other (i.e., the graph of $(x-2)^3$ over the interval $[3, 4]$ is the graph of x^3 over the interval $[1, 2]$ shifted to the right by 2 units).

41. $\int_1^4 f(x)\, dx = \int_1^2 f(x)\, dx + \int_2^4 f(x)\, dx = -1 + 7 = 6$

43. $\int_0^1 f(x)\, dx = \int_0^2 f(x)\, dx - \int_1^2 f(x)\, dx = 2 - (-1) = 3$

45. $\int_2^4 (f(x) - 2)\, dx = \int_2^4 f(x)\, dx - 2\int_2^4 dx = 7 - 2 \cdot 2 = 3$

47. Since the definite integral measures signed area, $-1 = \int_1^2 f(x)\, dx < 0$ implies that $f(x) < 0$ over some (or all) of the interval $[1, 2]$.

49. $\int_0^{\pi} \cos x\, dx = \int_0^{\pi/2} \cos x\, dx + \int_{\pi/2}^{\pi} \cos x\, dx = 0$ because the cosine function is symmetric about the point $(\pi/2, 0)$: $\int_{\pi/2}^{\pi} \cos x\, dx = -\int_0^{\pi/2} \cos x\, dx$.

    More formally: Since $\cos x$ is an even function, $\int_{-\pi/2}^{0} \cos x\, dx = \int_0^{\pi/2} \cos x\, dx$. Therefore,
    $$\int_0^{\pi} \cos x\, dx = \int_0^{\pi/2} \cos x\, dx + \int_{\pi/2}^{\pi} \cos x\, dx$$
    $$= \int_0^{\pi/2} \cos x\, dx + \int_0^{\pi/2} \left(-\cos(x - \pi/2)\right) dx$$
    $$= \int_0^{\pi/2} \cos x\, dx - \int_{-\pi/2}^{0} \cos x\, dx$$
    $$= 0.$$

51. The integrand is an odd function and the the interval of integration is symmetric about $x = 0$. Therefore, $\int_{-1}^{1} (4x^3 - 2x)\,dx = 0$.

53. Since $\sin(x + \pi) = -\sin x$ for all x,

$$\begin{aligned}\int_0^{2\pi} \sin x\,dx &= \int_0^{\pi} \sin x\,dx + \int_{\pi}^{2\pi} \sin x\,dx \\ &= \int_0^{\pi} \sin x\,dx + \int_0^{\pi} \sin(x + \pi)\,dx \\ &= \int_0^{\pi} \sin x\,dx - \int_0^{\pi} \sin x\,dx \\ &= 0\end{aligned}$$

55. $\int_0^{\pi} (1 + \sin x)\,dx = \int_0^{\pi} dx + \int_0^{\pi} \sin x\,dx = \pi + 2$

57. $\int_0^{\pi/2} (x + \cos x)\,dx = \int_0^{\pi/2} x\,dx + \int_0^{\pi/2} \cos x\,dx$. Now, $\int_0^{\pi/2} x\,dx = \frac{1}{2} \cdot \left(\frac{\pi}{2} - 0\right) \cdot \frac{\pi}{2} = \frac{\pi^2}{8}$ and $\int_0^{\pi/2} \cos x\,dx = \int_0^{\pi/2} \sin(x + \pi/2)\,dx = \int_{\pi/2}^{\pi} \sin x\,dx = \frac{1}{2} \int_0^{\pi} \sin x\,dx = \frac{1}{2} \cdot 2 = 1$. Therefore, $\int_0^{\pi/2} (x + \cos x)\,dx = \pi^2/8 + 1$.

59. Note that

$$\begin{aligned}\int_0^{2\pi} |\sin x|\,dx &= \int_0^{\pi} |\sin x|\,dx + \int_{\pi}^{2\pi} |\sin x|\,dx \\ &= \int_0^{\pi} \sin x\,dx - \int_{\pi}^{2\pi} \sin x\,dx \\ &= \int_0^{\pi} \sin x\,dx - \int_0^{\pi} \sin(x + \pi)\,dx \\ &= \int_0^{\pi} \sin x\,dx + \int_0^{\pi} \sin(x + \pi)\,dx \\ &= 2\int_0^{\pi} \sin x\,dx \\ &= 4\end{aligned}$$

Since the sine function has period 2π, it follows that $\int_0^{100\pi} |\sin x|\,dx = 50 \int_0^{2\pi} |\sin x|\,dx = 50 \cdot 4 = 200$.

61. $0.442 < \frac{5\pi}{12} - \frac{\sqrt{3}}{2} \le f(x) \le \frac{\pi}{12} + \frac{\sqrt{3}}{2} < 1.128$ if $0 \le x \le 3$. Therefore, $1.3 < 1.326 = 3 \cdot 0.442 \le \int_0^3 f(x)\,dx \le 3 \cdot 1.128 = 3.384 < 3.5$.

63. Since $-1 \le \cos x \le -1/2$ if $2\pi/3 \le x \le \pi$, $-\pi/3 = -1(\pi - 2\pi/3) \le \int_{2\pi/3}^{\pi} \cos x\,dx \le -\frac{1}{2}(\pi - 2\pi/3) = -\pi/6$

65. Any function f that is continous on the interval $[1, 5]$ and takes on both positive and negative values will have the desired property (e.g., $f(x) = x - 2$).

67. $\int_{-3}^{1} f(x)\,dx = 2 \cdot 4 = 8$ and $\int_{-3}^{7} f(x)\,dx = 10 \cdot 5 = 50$, so $\int_{1}^{7} f(x)\,dx = 42$. Therefore, the average value of f over the interval $[1, 7]$ is $42/6 = 7$.

69. (b) $\int_0^2 g(x)\,dx$ is the area of the region bounded above by the line $y = 2x$ and below by the x-axis between $x = 0$ and $x = 2$. Similarly, $\int_0^2 f(x)\,dx$ is the area of the region bounded above by the curve $y = x^2$ and below by the x-axis between $x = 0$ and $x = 2$. Since the line $y = 2x$ and the curve $y = x^2$ intersect at $x = 0$ and at $x = 2$, the area of the region bounded by these curves is the area that is below the line $y = 2x$ and above the curve $y = x^2$ between $x = 0$ and $x = 2$. Thus, the area of the region between the two curves is $\int_0^2 g(x)\,dx - \int_0^2 f(x)\,dx$.

71. (b) The curves $y = f(x)$ and $y = g(x)$ intersect at the point $x = 2$ and $0 \le g(x) \le f(x)$ if $0 \le x \le 2$. Therefore, the area of the region is $\int_0^2 \bigl(f(x) - g(x)\bigr)\,dx$.

73. (a) $\int_{-a}^{a} x^2\,dx$ is the area of the region bounded by the lines $x = -a$, $x = a$, and $y = 0$, and the curve $y = x^2$. The area of this region is the area of the rectangle bounded by the lines $x = -a$, $x = a$, and $y = 0$, and $y = a^2$ minus the area of the parabolic arch enclosed by the lines $x = -a$ and $x = a$, and the curve $y = x^2$: $2a^3 - 4a^3/3 = 2a^3/3$.

(b) $\dfrac{2a^3}{3} = \int_{-a}^{a} x^2\,dx = \int_{-a}^{0} x^2\,dx + \int_{0}^{a} x^2\,dx = 2\int_{0}^{a} x^2\,dx$, so $\int_{0}^{a} x^2\,dx = \dfrac{a^3}{3}$.

(c) $\int_{a}^{b} x^2\,dx = \int_{0}^{b} x^2\,dx - \int_{0}^{a} x^2\,dx = \dfrac{b^3}{3} - \dfrac{a^3}{3} = \dfrac{b^3 - a^3}{3}$.

75. HINTS: $bf(b)$ and $af(a)$ are the areas of rectangles. Just as $\int_{a}^{b} f(x)\,dx$ is the area of the region below the curve $y = f(x)$ and above the x-axis, $\int_{f(a)}^{f(b)} f^{-1}(y)\,dy$ is the area of the region bounded on the left by the y-axis, on the right by the curve $y = f(x)$, below by the line $y = f(a)$, and above by the line $y = f(b)$.

77. Let $f(x) = \sqrt{x}$. Then $f^{-1}(x) = x^2$ so

$$\int_{a}^{b} \sqrt{x}\,dx = b\sqrt{b} - a\sqrt{a} - \int_{\sqrt{a}}^{\sqrt{b}} x^2\,dx$$

$$= b^{3/2} - a^{3/2} - \int_{\sqrt{a}}^{\sqrt{b}} x^2\,dx$$

SECTION 5.2 THE AREA FUNCTION

§5.2 The Area Function

1. Suppose that $x < 0$. Then, $\int_x^0 t\,dt = -x^2/2$ because the value of the integral is the signed area of a right triangle with base $|x|$ and altitude $|x|$ that lies below the x-axis. Therefore, $\int_0^x t\,dt = -\int_x^0 t\,dt = x^2/2$.

3. $A_f(5) > A_f(1)$ because f is positive on the interval $[1, 5]$.

5. $A_f(-1) = \int_0^{-1} f(t)\,dt = -\int_{-1}^0 f(t)\,dt$. Now, $\int_{-1}^0 f(t)\,dt > 0$ since f is positive on the interval $[-1, 0]$, so $A_f(-1) < 0$.

7. Note that $A_f(-1) = \int_0^{-1} f(t)\,dt = -\int_{-1}^0 f(t)\,dt$ and
$$A_f(-2) = \int_0^{-2} f(t)\,dt = -\int_{-2}^0 f(t)\,dt = -\left(\int_{-2}^{-1} f(t)\,dt + \int_{-1}^0 f(t)\,dt\right) = -\int_{-2}^{-1} f(t)\,dt + A_f(-1).$$
Now, $\int_{-2}^{-1} f(t)\,dt > 0$ since f is positive on the interval $[-2, -1]$ so $A_f(-1) > A_f(-2)$.

9. Note that $F(7) = \int_0^7 f(t)\,dt = \int_0^1 f(t)\,dt + \int_1^7 f(t)\,dt = F(1) + \int_1^7 f(t)\,dt$. Now, since $f(t) > 1$ on the interval $[1, 7]$, $\int_1^7 f(t)\,dt > 6$ so $F(7) - F(1) > 6 \implies F(7) > F(1)$.

11. Since $f(t) > 1 \geq 0$ if $t \geq 0$, $F(x) = \int_0^x f(t)\,dt > 0$ for all $x > 0$. Furthermore, since $F(0) = 0$, $F(x) > F(0)$ for all $x > 0$.

13. $F(x) = \int_0^x f(t)\,dt > \int_0^x 1\,dt = x$ if $x > 0$. Therefore, $F(6) \geq 6$.

15. $F(6) = \int_0^6 f(t)\,dt = \int_0^2 f(t)\,dt + \int_2^6 f(t)\,dt = F(2) + \int_2^6 f(t)\,dt$. Since $f(t) < 0$ for all $t \geq 0$, $\int_2^6 f(t)\,dt < 0$. This implies that $F(6) < F(2)$.

17. For any $x > 1$, $F(x) = \int_0^x f(t)\,dt = \int_0^1 f(t)\,dt + \int_1^x f(t)\,dt = F(1) + \int_1^x f(t)\,dt$. Since $f(t) < 0$ for all $t \geq 0$, $\int_1^x f(t)\,dt < 0$. This implies that $F(x) < F(1)$ for all $x > 1$.

19. No, because $F(4) = \int_0^4 f(t)\,dt < \int_0^4 (-2)\,dt = -8$.

21. Using the formula for the area of a trapezoid: $F(x) = \int_0^x (3t+2)\,dt = \frac{1}{2}x(2 + 3x + 2) = 3x^2/2 + 2x$, and $G(x) = \int_1^x (3t+2)\,dt = \frac{1}{2}(x-1)(5 + 3x + 2) = 3x^2/2 + 2x - 7/2$. Using the formula for the area of a triangle: $H(x) = \int_{-2}^x (3t+2)\,dt = \int_{-2}^{-2/3}(3t+2)\,dt + \int_{-2/3}^x (3t+2)\,dt = -\frac{1}{2} \cdot 4 \cdot \frac{4}{3} + \frac{1}{2} \cdot (3x+2) \cdot (x+2/3) = 3x^2/2 + 2x - 2$.

23. $F(x) - H(x) = \int_0^x f(t)\,dt - \int_{-2}^x f(t)\,dt = \int_0^x f(t)\,dt - \int_{-2}^0 f(t)\,dt - \int_0^x f(t)\,dt$
$= -\int_{-2}^0 f(t)\,dt = -H(0) = 2$

25. Since $F'(x) = 3x + 2$, F is decreasing on the interval $(-\infty, -2/3)$. $f(x) < 0$ on this interval.

27. No. Since $F''(x) = 3 > 0$, F is concave up on the interval $(-\infty, \infty)$.

29. Using the formula for the area of a trapezoid: $F(x) = \int_0^x (2 - 3t)\,dt = \frac{1}{2}x(2 + 2 - 3x) = 2x - 3x^2/2$, and $G(x) = \int_1^x (2 - 3t)\,dt = -\frac{1}{2}(x-1)(1+3x-2) = 2x - 3x^2/2 - 1/2$. Using the formula for the area of a triangle:
$H(x) = \int_{-2}^x (3t+2)\,dt = \int_{-2}^{2/3}(2-3t)\,dt + \int_{2/3}^x (2-3t)\,dt = \frac{1}{2}\cdot 8 \cdot \frac{8}{3} - \frac{1}{2}\cdot(3x-2)\cdot(x-2/3) = 2x - 3x^2/2 + 10$.

31. $F(x) - H(x) = F(x) - \big(H(0) + F(x)\big) = -H(0) = -10$

33. $F'(x) = 2 - 3x$, so F is decreasing on the interval $(2/3, \infty)$. $f(x) < 0$ on this interval.

35. $F''(x) = -3$, so F is concave down on the interval $(-\infty, \infty)$. f is decreasing on this interval.

37. (a) Since $F' = f$, F is decreasing on $[a, b]$.
    (b) Since $F'' = f'$, F is concave up on $[a, b]$.

39. $f(x) = \begin{cases} 2+x, & x \le 0 \\ 2-x, & x > 0 \end{cases}$. Thus, using the geometric formulas for the areas of triangles and trapezoids,

$A_f(x) = \int_0^x f(t)\,dt = \begin{cases} 2x + x^2/2, & x \le 0 \\ 2x - x^2/2, & x > 0. \end{cases}$

41. A_f is decreasing on the intervals $(-\infty, -2)$ and $(2, \infty)$. f is negative on these intervals.

43. A_f is concave down on the interval $(0, \infty)$. f is decreasing on this interval.

45. Note that $\int_0^x f(t)\,dt = \int_0^3 f(t)\,dt + \int_3^x f(t)\,dt$ so $A_f(x) = \int_3^x f(t)\,dt - \int_0^3 f(t)\,dt$. Since $\int_0^3 f(t)\,dt = -3/2$, $A_f(x) = \begin{cases} 2x + x^2/2 - 3/2, & x \le 0 \\ 2x - x^2/2 - 3/2, & x > 0. \end{cases}$

47. A_f is decreasing over the intervals $(-\infty, -2)$ and $(2, \infty)$. f is negative on these intervals.

49. A_f is concave down on the interval $(0, \infty)$. f is decreasing on this interval.

51. $G(x) = \int_{-3}^x f(t)\,dt = \int_{-3}^2 f(t)\,dt + \int_2^x f(t)\,dt = \int_{-3}^2 f(t)\,dt + H(x) = H(x) - 2\pi + 1/2$. The value of G includes the signed area of the region bounded by f over the interval $[-3, 2]$ but the value of H does not.

53. At $x = 1$, the sign of f changes from negative to positive. Thus, to the left of $x = 1$ the signed area is negative while to the right it is positive.

55. The extreme values of G over the interval $[-5, -2]$ must occur at an endpoint of the interval or at $x = -3$ (since f changes sign there). Since $G(-3) = 0$, $G(-5) = -3$, and $G(-2) > -2$, G achieves its minimum value at $x = -5$ and its maximum value at $x = -3$. $G(-5) = -3$ and $G(-3) = 0$.

57. $\int_0^{\sqrt{\pi/2}} f(t)\,dt = \sin\big((\sqrt{\pi/2})^2\big) = \sin(\pi/2) = 1$

59. $\int_{-\sqrt{3\pi/2}}^x f(t)\,dt = \int_{-\sqrt{3\pi/2}}^0 f(t)\,dt + \int_0^x f(t)\,dt = -\int_0^{-\sqrt{3\pi/2}} f(t)\,dt + \int_0^x f(t)\,dt$
$= -\sin(3\pi/2) + \sin(x^2) = 1 + \sin(x^2)$

61. $\int_{-1}^4 \sqrt{1+u^4}\,du = F(4)$

SECTION 5.2 THE AREA FUNCTION

63. $\int_{-2}^{3} \sqrt{1+t^4}\, dt = \int_{-2}^{-1} \sqrt{1+t^4}\, dt + \int_{-1}^{3} \sqrt{1+t^4}\, dt = -\int_{-1}^{-2} \sqrt{1+t^4}\, dt + F(3) = F(3) - F(-2)$

65. $G(x) - F(x) = \int_{a}^{x} g(t)\, dt - \int_{a}^{x} f(t)\, dt = \int_{a}^{x} \big(g(t) - f(t)\big)\, dt \geq 0$ since $g(t) - f(t) \geq 0$ for all $t \geq a$.

67. (a) When $x \geq 0$, $\int_{0}^{x} \sqrt{1-t^2}\, dt$ can be evaluated by adding together the area of the circular sector with angle θ and the area of the right triangle shown in the diagram. Since $\theta = \arcsin x$, the areas are $\frac{1}{2}\arcsin x$ and $\frac{1}{2} x \sqrt{1-x^2}$, respectively.

(b) A justification similar to the one in part (a) can be given for the $x < 0$ case.

(c) $\left(\frac{1}{2} x \sqrt{1-x^2} + \frac{1}{2} \arcsin x \right)' = \frac{-x^2}{2\sqrt{1-x^2}} + \frac{\sqrt{1-x^2}}{2} + \frac{1}{2\sqrt{1-x^2}} = \sqrt{1-x^2}$

69. $\int_{0}^{x} \sqrt{a^2 - t^2}\, dt = \frac{1}{2} x \sqrt{a^2 - x^2} + \frac{a^2}{2} \arcsin(x/a)$

71. (a) $A_f(x) = x^3/3 + 9$

(b) Yes.

§5.3 The Fundamental Theorem of Calculus

1. Yes, because $F'(3) = f(3) > 0$.

3. Yes, because f is increasing at $x = 3$.

5. Yes, $F'(4) = f(4) > 20$.

7. Since $f(x) > 10$ if $0 \le x \le 2$, $F(2) = \int_0^2 f(t)\,dt > \int_0^2 10\,dt = 20$

9. F has stationary points where $F'(x) = f(x) = 0$. Thus, F has stationary points at $x = -3.5$, $x = -2$, and $x = 2$.

11. F is decreasing on the intervals where $F'(x) = f(x) < 0$. Thus, F is decreasing on the intervals $[-5, -3.5)$ and $(2, 5]$.

13. $F(0) = \int_0^0 f(t)\,dt = 0$, $F'(0) = f(0) = 2$, and $F''(0) = f'(0) = 1$.

15. Since $F(-1) = -1.5$ and $F'(-1) = 1$, the tangent line has slope 1 and passes through the point $(-1, -1.5)$. Therefore, $y = (x+1) - 1.5 = x - 0.5$ is an equation of the tangent line.

17. F has an inflection point where $F''(x) = f'(x)$ changes sign. Thus, F has an inflection point at $x = -3$, $x = -2$, $x = 1$, and $x = 3$.

19. $\int_1^3 x\,dx = \dfrac{1}{2}x^2 \Big]_1^3 = \dfrac{9}{2} - \dfrac{1}{2} = 4$

21. $\int_{-1}^1 x^2\,dx = \dfrac{x^3}{3}\Big]_{-1}^1 = \dfrac{1}{3} - \dfrac{-1}{3} = \dfrac{2}{3}$

23. $\int_{-1}^2 e^x\,dx = e^x\big]_{-1}^2 = e^2 - e^{-1}$

25. $\int_{\pi/2}^{\pi} \sin x\,dx = -\cos x\big]_{\pi/2}^{\pi} = -\cos\pi + \cos(\pi/2) = 1$

27. $\int_1^4 (5x + x^{3/2})\,dx = \dfrac{5x^2}{2} + \dfrac{x^{5/2}}{5/2}\Big]_1^4 = 499/10$

29. $\int_{-\pi/2}^{\pi} \sin(2x)\,dx = -\dfrac{1}{2}\cos(2x)\Big]_{-\pi/2}^{\pi} = -\dfrac{1}{2}\cos(2\pi) + \dfrac{1}{2}\cos\pi = -1$

31. (a) By the fundamental theorem, $F'(x) = \sqrt[3]{x^2 + 7}$, so $F'(1) = \sqrt[3]{8} = 2$. Thus the slope of the tangent line is 2; since the graph of F passes through the point $(1, 0)$, $y = 2x - 2$ is an equation for the tangent line.

    (b) No. Since $F''(x) > 0$ on the interval $[1, 1.1]$, the graph of F is concave up over this interval. This implies that the tangent line estimate of $F(1.1)$ is too small.

33. (a) Using the chain rule, $\big(\sin(x^2)\big)' = 2x\cos(x^2)$.

    (b) The result in part (a) implies that $\frac{1}{2}\sin(x^2)$ is an antiderivative of $x\cos(x^2)$. Therefore,
    $$\int_0^3 x\cos(x^2)\,dx = \dfrac{1}{2}\sin(x^2)\Big]_0^3 = \dfrac{1}{2}(\sin 9 - \sin 0) = \dfrac{1}{2}\sin 9.$$

SECTION 5.3 THE FUNDAMENTAL THEOREM OF CALCULUS

35. Yes, F has a local maximum at points where $f(x) = 0$ and f changes sign from positive to negative. Thus, F has local maxima at $x = 1$, $x = 5$, and $x = 9$.

37. F has local minima at $x = 3$ and $x = 7$, but $F(3) < F(7) < F(10)$ since $\int_3^7 f(t)\,dt > 0$ and $\int_7^{10} f(t)\,dt > 0$. Furthermore, $F(3) < F(1)$ since $\int_1^3 f(t)\,dt < 0$. Therefore, F attains its minimum value on the interval $[1, 10]$ at $x = 3$.

39. Yes, $F'(2) = f(2) < -0.6 < -0.4$.

41. No, $F''(7) = f'(7) \approx 0.2$.

43. Let $F(x) = \int_0^x f(t)\,dt = 3x^2 + e^x - \cos x$. Since $F'(x) = f(x) = 6x + e^x + \sin x$, $f(2) = 12 + e^2 + \sin 2$.

45. Since $\left(\int_1^x \sqrt[3]{1+t^2}\,dt\right)' = \sqrt[3]{1+t^2}$ and $\int_1^1 \sqrt[3]{1+t^2}\,dt = 0$, $f(x) = \int_1^x \sqrt[3]{1+t^2}\,dt$ is the unique solution of the IVP $f'(x) = \sqrt[3]{1+x^2}$, $f(1) = 0$.

47. According to the fundamental theorem, $g'(x) = f(x)$. Furthermore, $g(a) = b + \int_a^a f(t)\,dt = b$. Therefore, g is a solution of the IVP.

49. Using the chain rule, $\dfrac{d}{dx}(F(x^2)) = F'(x^2) \cdot 2x = 2x\sqrt[3]{1+x^4}$ since $F'(x) = \sqrt[3]{1+x^2}$.

51. The average value of f' over the interval $[1, 10]$ is $\dfrac{\int_1^{10} f'(x)\,dx}{10 - 1} = \dfrac{f(10) - f(1)}{9} = \dfrac{7 - 13}{9} = -\dfrac{2}{3}$.

53. The area of the region enclosed by one arch of the sine function is
$$\int_0^\pi \sin x\,dx = -\cos x\Big]_0^\pi = -\cos \pi + \cos 0 = 2.$$

55. $g'(4) = f(4) = 2$.

57. g is concave up where $g'' = f'$ is positive. Thus, g is concave up on the interval $(1, 4)$.

59. $g'(4) = f(4) = 4$

61. The properties listed imply that f changes sign from positive to negative in the interval $[0, 7]$ only at the point $x = 5$. Therefore, g must achieve its maximum value on the interval $[0, 7]$ at this point.

63. The average value of f' over the interval $[2, 5]$ is $\dfrac{\int_2^5 f'(x)\,dx}{5 - 2} = \dfrac{f(5) - f(2)}{3} = -\dfrac{4}{3}$.

65. (a) The demand at time t is $D(t) = 800 - 10t$ where t measures time in months. The production at time t is $P(t) = 900$.

(b) $D(t)$ represents the demand for the product in units/month. Since $D(t)$ is a rate function, $\int_0^t D(s)\,ds$ is the accumulated demand over the interval $[0, t]$.

(c) The inventory at time t is $I(t) =$ starting inventory $+$ total production up to time t $-$ total demand up to time t. Since $P(t) = 900 - Rt$ (where R is the rate at which production is decreased measured in units per month),
$$I(t) = 1680 + \int_0^t (900 - Rs)\,ds - \int_0^t (800 - 10s)\,ds$$
$$= (5 - \frac{R}{2})t^2 + 100t + 1680.$$

(d) When $t = 12$ the inventory is $3600 - 72R$; to make this zero, $R = 50$ must be true.

§5.4 Finding antiderivatives; the method of substitution

1. $u = 4x + 3 \implies du = 4\,dx$ so $\int (4x+3)^{-3}\,dx = \int u^{-3} \cdot \dfrac{du}{4} = \dfrac{1}{4} \cdot \dfrac{-1}{2} u^{-2} + C = -\dfrac{1}{8(4x+3)^2} + C$

3. $u = \sin x \implies du = \cos x\,dx$ so $\int e^{\sin x} \cos x\,dx = \int e^u\,du = e^u + C = e^{\sin x} + C$.

5. $u = \arctan x \implies du = (1+x^2)^{-1}\,dx$ so $\int \dfrac{\arctan x}{1+x^2}\,dx = \int u\,du = \dfrac{1}{2} u^2 + C = \dfrac{1}{2}(\arctan x)^2 + C$.

7. $u = x^{-1} \implies du = -x^{-2}\,dx$ so $\int \dfrac{e^{1/x}}{x^2}\,dx = -\int e^u\,du = -e^u + C = -e^{x^{-1}} + C = -e^{1/x} + C$.

9. $u = e^x \implies du = e^x\,dx$ so $\int \dfrac{e^x}{1+e^{2x}}\,dx = \int \dfrac{du}{1+u^2} = \arctan u + C = \arctan(e^x) + C$.

11. Let $u = x^2$. Then, $a = (-2)^2 = 4$, $b = 1^2 = 1$, and $du = 2x\,dx$ so
$$\int_{-2}^{1} \dfrac{x}{1+x^4}\,dx = \dfrac{1}{2}\int_{4}^{1} \dfrac{du}{1+u^2} = \dfrac{1}{2}\arctan u\Big]_{4}^{1} = (\pi/4 - \arctan 4)/2 \approx -0.27021.$$

13. Let $u = 2x^2 + 1$. Then, $a = 2 \cdot (0)^2 + 1 = 1$, $b = 2 \cdot (3)^2 + 1 = 19$, and $du = 4x\,dx$, so
$$\int_{0}^{3} \dfrac{x}{(2x^2+1)^3}\,dx = \dfrac{1}{4}\int_{1}^{19} u^{-3}\,du = -\dfrac{1}{8u^2}\Big]_{1}^{19} = \dfrac{45}{361}.$$

15. Using the chain rule, $\left(\dfrac{1}{2}\arctan(x^2) + C\right)' = \dfrac{1}{2}\dfrac{2x}{1+(x^2)^2} = \dfrac{x}{1+x^4}$.

17. If $u = 1 + \sqrt{x}$, then $du = \frac{1}{2}x^{-1/2}\,dx$ and $2(u-1)\,du = dx$. Thus,
$$\int \dfrac{dx}{1+\sqrt{x}} = \int \dfrac{2(u-1)\,du}{u} = 2u - 2\ln|u| + C = 2(1+\sqrt{x}) - 2\ln(1+\sqrt{x}) + C.$$

19. If $u = a + b/x$, then $du = -\dfrac{b}{x^2}\,dx$. Therefore,
$$\int \dfrac{dx}{ax^2+bx} = \int \dfrac{dx}{x^2(a+b/x)} = -\dfrac{1}{b}\int \dfrac{du}{u} = -\dfrac{1}{b}\ln|u| + C = -\dfrac{1}{b}\ln|a+b/x| + C = \dfrac{1}{b}\ln\left|\dfrac{x}{ax+b}\right| + C.$$

21. Make the substitutions $u = 3x$ and $du = 3\,dx$ in the *second integral*:
$$\int_{0}^{4} g(3x)\,dx = \dfrac{1}{3}\int_{0}^{12} g(u)\,du = \dfrac{1}{3}\pi.$$

23. Let $u = 2x + 3$. Then, $du = 2\,dx$ so
$$\int \cos(2x+3)\,dx = \dfrac{1}{2}\int \cos u\,du = \dfrac{1}{2}\sin u + C = \dfrac{1}{2}\sin(2x+3) + C.$$

25. Let $u = 3x - 2$. Then $du = 3\,dx$ so $\int (3x-2)^4\,dx = \dfrac{1}{3}\int u^4\,du = \tfrac{1}{15}u^5 + C = \tfrac{1}{15}(3x-2)^5 + C$.

27. Let $u = 1 + x^4$. Then, $du = 4x^3\,dx$ so $\int \dfrac{2x^3}{1+x^4}\,dx = \dfrac{1}{2}\int \dfrac{du}{u} = \dfrac{1}{2}\ln|u| + C = \dfrac{1}{2}\ln(1+x^4) + C$.

29. Let $u = 1 - 2x$. Then, $du = -2\,dx$ to $\int \dfrac{dx}{1-2x} = -\dfrac{1}{2}\int \dfrac{du}{u} = -\tfrac{1}{2}\ln|u| + C = -\tfrac{1}{2}\ln|1-2x| + C$.

SECTION 5.4 FINDING ANTIDERIVATIVES; THE METHOD OF SUBSTITUTION

31. Let $u = 3 - 2x$. Then, $x = (3-u)/2$ and $du = -2\,dx$ so

$$\int x\sqrt{3-2x}\,dx = -\frac{1}{4}\int (3-u)u^{1/2}\,du$$
$$= -\frac{1}{4}\int \left(3u^{1/2} - u^{3/2}\right)du$$
$$= \frac{1}{10}u^{5/2} - \frac{1}{2}u^{3/2} + C$$
$$= \frac{1}{10}(3-2x)^{5/2} - \frac{1}{2}(3-2x)^{3/2} + C.$$

33. Let $u = 1 + x^{-1}$. Then, $du = -x^{-2}$ so

$$\int \frac{\sqrt{1+x^{-1}}}{x^2}\,dx = -\int \sqrt{u}\,du = -\frac{2}{3}u^{3/2} + C = -\frac{2}{3}\left(1+x^{-1}\right)^{3/2} + C.$$

35. Let $u = x^4 - 1$. Then, $du = 4x^3\,dx$ so

$$\int x^3\left(x^4-1\right)^2 dx = \frac{1}{4}\int u^2\,du = \frac{1}{12}u^3 + C = \frac{1}{12}\left(x^4-1\right)^3 + C.$$

37. Let $u = x^3$. Then, $du = 3x^2\,dx$ so $\displaystyle\int \frac{x^2}{1+x^6}\,dx = \frac{1}{3}\int \frac{du}{1+u^2} = \frac{1}{3}\arctan u + C = \frac{1}{3}\arctan\left(x^3\right) + C.$

39. Let $u = 4x^3 + 5$. Then, $du = 12x^2\,dx$ so

$$\int x^2\sqrt{4x^3+5}\,dx = \frac{1}{12}\int \sqrt{u}\,du = \frac{1}{18}u^{3/2} + C = \frac{1}{18}\left(4x^3+5\right)^{3/2} + C.$$

41.
$$\int \frac{x+4}{x^2+1}\,dx = \int \frac{x}{x^2+1}\,dx + 4\int \frac{dx}{x^2+1} = \int \frac{x}{x^2+1}\,dx + 4\arctan x = \frac{1}{2}\ln\left(1+x^2\right) + 4\arctan x + C.$$
(The substitution $u = 1 + x^2$ was used in the last step.)

43. Let $u = x^2 + 3x + 5$. Then, $du = (2x+3)\,dx$ so

$$\int \frac{2x+3}{\left(x^2+3x+5\right)^4}\,dx = \int \frac{du}{u^4} = -\frac{1}{3}u^{-3} + C = -\frac{1}{3}\left(x^2+3x+5\right)^{-3} + C.$$

45. Let $u = 3x^2 + 6x + 5$. Then, $du = 6(x+1)\,dx$ so

$$\int \frac{x+1}{\sqrt[3]{3x^2+6x+5}}\,dx = \frac{1}{6}\int \frac{du}{u^{1/3}} = \frac{1}{4}u^{2/3} + C = \frac{1}{4}\left(3x^2+6x+5\right)^{2/3} + C.$$

47. Let $u = 2e^x + 3$. Then, $du = 2e^x\,dx$ so

$$\int \frac{e^x}{(2e^x+3)^2}\,dx = \frac{1}{2}\int u^{-2}\,du = -\frac{1}{2}u^{-1} + C = -\frac{1}{2}(2e^x+3)^{-1} + C.$$

49. Let $u = x + 1$. Then, $du = dx$ so

$$\int \frac{2x+3}{(x+1)^2}\,dx = \int \frac{2u+1}{u^2}\,du = 2\int u^{-1}\,du + \int u^{-2}\,du$$
$$= 2\ln|u| - u^{-1} + C = 2\ln|x+1| - (x+1)^{-1} + C.$$

51. Let $u = \cos x$. Then, $du = -\sin x\,dx$ so

$$\int \tan x\,dx = \int \frac{\sin x}{\cos x}\,dx = -\int \frac{du}{u} = -\ln|u| + C = -\ln|\cos x| + C.$$

53. Let $u = \arcsin x$. Then, $du = (1-x^2)^{-1/2}\,dx$ so

$$\int \frac{\arcsin x}{\sqrt{1-x^2}}\,dx = \int u\,du = \frac{1}{2}u^2 + C = \frac{1}{2}(\arcsin x)^2 + C.$$

55. Let $u = 3x^2 + 4$. Then, $du = 6x\,dx$ so $\int \dfrac{5x}{3x^2+4}\,dx = \dfrac{5}{6}\int \dfrac{du}{u} = \ln|u| + C = \dfrac{5}{6}\ln\left(3x^2+4\right) + C$.

57. Let $u = \tan x$. Then, $du = \sec^2 x\,dx$ so
$$\int \frac{\sec^2 x}{\sqrt{1-\tan^2 x}}\,dx = \int \frac{du}{\sqrt{1-u^2}} = \arcsin u + C = \arcsin(\tan x) + C.$$

59. Let $u = x^2$. Then, $du = 2x\,dx$ so
$$\int x\tan\left(x^2\right)\,dx = \frac{1}{2}\int \tan u\,du = \frac{1}{2}\int \frac{\sin u}{\cos u}\,du = -\frac{1}{2}\ln|\cos u| + C = -\frac{1}{2}\ln\left|\cos\left(x^2\right)\right| + C.$$

61. Let $u = \sin x$. Then, $du = \cos x\,dx$ so
$$\int \frac{\cos x}{\sin^4 x}\,dx = \int \frac{du}{u^4} = \frac{1}{3u^3} + C = -\frac{1}{3\sin^3 x} + C = -\tfrac{1}{3}\csc^3 x + C.$$

63. Let $u = 1 + \sqrt{x}$. Then $du = \tfrac{1}{2}x^{-1/2}\,dx$ so
$$\int \frac{\left(1+\sqrt{x}\right)^3}{\sqrt{x}}\,dx = 2\int u^3\,du = \tfrac{1}{2}u^4 + C = \tfrac{1}{2}\left(1+\sqrt{x}\right)^4 + C.$$

65. Let $u = x^2 + 2$. Then $x^2 = u - 2$ and $du = 2x\,dx$, so
$$\int x^3\sqrt{x^2+2}\,dx = \int x^2\sqrt{x^2+2}\cdot x\,dx$$
$$= \frac{1}{2}\int (u-2)u^{1/2}\,du = \frac{1}{2}\int \left(u^{3/2} - 2u^{1/2}\right)\,du$$
$$= \tfrac{1}{5}u^{5/2} - \tfrac{2}{3}u^{3/2} + C$$
$$= \tfrac{1}{5}(x^2+2)^{5/2} - \tfrac{2}{3}(x^2+2)^{3/2} + C.$$

67. Let $u = e^x$. Then, $du = e^x\,dx$ so
$$\int \frac{e^x}{e^{2x} + 2e^x + 1}\,dx = \int \frac{du}{u^2+2u+1} = \int \frac{du}{(u+1)^2}$$
$$= -(u+1)^{-1} + C$$
$$= -(e^x+1)^{-1} + C = -\frac{1}{e^x+1} + C.$$

69. Let $u = 1 + x^2$. Then, $du = 2x\,dx$ so $\displaystyle\int_0^2 \frac{x}{(1+x^2)^3}\,dx = \frac{1}{2}\int_1^5 \frac{du}{u^3} = -\frac{1}{4}u^{-2}\Big]_1^5 = \frac{1}{4} - \frac{1}{100} = \frac{6}{25}$.

71. Let $u = \ln x$. Then, $du = x^{-1}\,dx$ so $\displaystyle\int_1^e \frac{\sin(\ln x)}{x}\,dx = \int_0^1 \sin u\,du = -\cos u\Big]_0^1 = 1 - \cos 1$.

73. Let $u = \sin x$. Then, $du = \cos x\,dx$ so $\displaystyle\int_0^\pi \sin^3 x \cos x\,dx = \int_0^0 u^3\,du = 0$.

SECTION 5.4 FINDING ANTIDERIVATIVES; THE METHOD OF SUBSTITUTION

75. Let $u = \pi - x$. Then $x = \pi - u$ and $du = -dx$. Therefore,

$$\int_0^\pi xf(\sin x)\,dx = -\int_\pi^0 (\pi - u)f\bigl(\sin(\pi - u)\bigr)\,du$$
$$= \int_0^\pi (\pi - u)f\bigl(\sin(\pi - u)\bigr)\,du$$
$$= \int_0^\pi (\pi - u)f(\sin u)\,du$$
$$= \pi \int_0^\pi f(\sin u)\,du - \int_0^\pi u f(\sin u)\,du$$

This implies that $2\int_0^\pi xf(\sin x)\,dx = \pi \int_0^\pi f(\sin x)\,dx$ or, equivalently, that
$\int_0^\pi xf(\sin x)\,dx = \dfrac{\pi}{2}\int_0^\pi f(\sin x)\,dx$.

77. First, note that $\displaystyle\int \frac{dx}{\sqrt{x} + \sqrt[3]{x}} = \int \frac{dx}{\sqrt[3]{x}\left(\sqrt[6]{x} + 1\right)}$. Now, let $u^6 = x$ and $w = u + 1$ to obtain

$$\int \frac{dx}{\sqrt{x} + \sqrt[3]{x}} = 6\int \frac{u^5}{u^2(u+1)}\,du = 6\int \frac{u^3}{u+1}\,du$$
$$= 6\int \frac{(w-1)^3}{w}\,dw = 6\int \left(w^2 - 3w + 3 - w^{-1}\right)\,dw$$
$$= 2w^3 - 9w^2 + 18w - 6\ln w + K$$
$$= 2x^{1/2} - 3x^{1/3} + 6x^{1/6} - 6\ln\left(x^{1/6} + 1\right) + C.$$

§5.5 Integral Aids: Tables and Computers

1. $\int \dfrac{dx}{3+2e^{5x}} = \dfrac{1}{3}x - \dfrac{1}{15}\ln\left(3+2e^{5x}\right) + C.$ [Use formula #58.]

3. $\int \dfrac{dx}{x^2(3-x)} = -\dfrac{1}{3x} - \dfrac{1}{9}\ln\left|\dfrac{3-x}{x}\right| = \dfrac{1}{9}\ln\left|\dfrac{x}{3-x}\right| - \dfrac{1}{3x} + C.$ [Use formula #24.]

5. $\int \dfrac{dx}{x\sqrt{2x+1}} = \ln\left|\dfrac{\sqrt{2x+1}-1}{\sqrt{2x+1}+1}\right| + C.$ [Use formula #28.]

7. $\int e^{2x}\cos(3x)\,dx = \dfrac{e^{2x}}{13}\left(2\cos(3x) + 3\sin(3x)\right) + C.$ [Use formula #55.]

9. $\int \dfrac{dx}{x\sqrt{3x-2}} = \dfrac{2}{\sqrt{2}}\arctan\sqrt{\dfrac{3x-2}{2}} + C.$ [Use formula #29.]

11. $\int \dfrac{x^2}{1+x^2}\,dx = \int\left(1 - \dfrac{1}{1+x^2}\right)dx = x - \arctan x + C$

13. $\int \dfrac{x-1}{x+1}\,dx = \int \dfrac{(x+1)-2}{x+1}\,dx = \int\left(1 - \dfrac{2}{x+1}\right)dx = x - 2\ln|x+1| + C$

15. $\int \dfrac{dx}{1+\sin x} = \int \dfrac{1-\sin x}{(1+\sin x)(1-\sin x)} = \int \dfrac{1-\sin x}{1-\sin^2 x} = \int \dfrac{1-\sin x}{\cos^2 x}\,dx$
$= \int\left(\sec^2 x - \sec x\tan x\right)dx = \tan x - \sec x + C$

17. Use the substitution $u = 3x$: $\int \dfrac{dx}{1+9x^2} = \int \dfrac{dx}{1+(3x)^2} = \dfrac{1}{3}\int \dfrac{du}{1+u^2}$
$= \dfrac{1}{3}\arctan u = \dfrac{1}{3}\arctan(3x) + C.$

19. $\int \tan^3(5x)\,dx = \dfrac{1}{10}\tan^2(5x) + \dfrac{1}{5}\ln|\cos(5x)| + C.$ [Use formulas #50 and #7.]

21. $\int \dfrac{2x+3}{4x+5}\,dx = 2\int \dfrac{x}{4x+5}\,dx + 3\int \dfrac{dx}{4x+5}$
$= 2\left(\dfrac{x}{4} - \dfrac{5}{16}\ln|4x+5|\right) + \dfrac{3}{4}\ln|4x+5|$
$= \dfrac{x}{2} + \dfrac{1}{8}\ln|4x+5| + C.$
[Use formulas #21 and #20.]

23. $\int \dfrac{4x+5}{(2x+3)^2}\,dx = 4\int \dfrac{x}{(2x+3)^2}\,dx + 5\int \dfrac{dx}{(2x+3)^2} = \dfrac{1}{2(2x+3)} + \ln|2x+3| + C.$ [Use formulas #22 and #19.]

25. $\int \dfrac{x+2}{2+x^2}\,dx = \int \dfrac{x}{2+x^2}\,dx + 2\int \dfrac{dx}{2+x^2} = \dfrac{1}{2}\ln\left(x^2+2\right) + \sqrt{2}\arctan(x\sqrt{2}/2) + C.$
[Use formulas #35 and #31.]

27. $\int \dfrac{5}{4x^2+20x+16}\,dx = \dfrac{5}{4}\int \dfrac{dx}{x^2+5x+4} = \dfrac{5}{4}\int \dfrac{dx}{(x+5/2)^2 - 9/4}\,dx = \dfrac{5}{12}\ln\left|\dfrac{x+1}{x+4}\right| + C.$
[Use formula #32.]

SECTION 5.5 INTEGRAL AIDS: TABLES AND COMPUTERS

29. Making the substitution $u = x^2$ together with formula #47 (or #49),

$$\int x^3 \cos(x^2)\, dx = \frac{1}{2} \int u \cos u\, du = \frac{1}{2}(u \sin u + \cos u) = \frac{1}{2}\left(x^2 \sin\left(x^2\right) + \cos\left(x^2\right)\right) + C$$

31. $\displaystyle\int \frac{dx}{\left(x^2 + 3x + 2\right)^2} = \int \frac{dx}{\left((x+3/2)^2 - 1/4\right)^2} = -\frac{2x}{x^2 + 3x + 2} - 2\ln\left|\frac{x+1}{x+2}\right| + C.$
[Use formulas #33 and #32.]

33. $\displaystyle\int \frac{e^x}{e^{2x} - 2e^x + 5}\, dx = \int \frac{du}{u^2 - 2u + 5} = \int \frac{du}{(u-1)^2 + 4} = \int \frac{dw}{w^2 + 4} = \frac{1}{2}\arctan(w/2) = \frac{1}{2}\arctan\left(\frac{e^x - 1}{2}\right) + C.$
[Use the substitution $u = e^x$ and formulas #13 or #31.]

35. $\displaystyle\int \sqrt{x^2 + 4x + 1}\, dx = \int \sqrt{(x+2)^2 - 3}\, dx$
$= \frac{1}{2}\left((x+2)\sqrt{x^2 + 4x + 1} - 3\ln\left|x + 2 + \sqrt{x^2 + 4x + 1}\right|\right) + C.$
[Complete the square, then use formula #36.]

37. $\displaystyle\int \frac{\cos x \sin x}{(\cos x - 4)(3\cos x + 1)}\, dx = \int \frac{\cos x \sin x}{3\cos^2 x - 11\cos x - 4}\, dx$
$= -\int \frac{u}{3u^2 - 11u - 4}\, du$
$= -\frac{1}{3}\int \frac{u}{u^2 - 11u/3 - 4/3}\, du = -\frac{1}{3}\int \frac{u}{(u - 11/6)^2 - 169/36}\, du$
$= -\frac{1}{3}\int \frac{w + 11/6}{w^2 - (13/6)^2}\, dw$
$= -\frac{4}{13}\ln|w - 13/6| - \frac{1}{39}\ln|w + 13/6|$
$= -\frac{4}{13}\ln|u - 4| - \frac{1}{39}\ln|u + 1/3|$
$= -\frac{4}{13}\ln|\cos x - 4| - \frac{1}{39}\ln|\cos x + 1/3| + C$
[Use the substitution $u = \cos x$ and formulas #35 and #32.]

39. $\displaystyle\int x \sin(3x + 4)\, dx = \frac{1}{9}\int (u - 4)\sin u\, du$
$= \frac{1}{9}(\sin u - u\cos u + 4\cos u)$
$= \frac{1}{9}(\sin(3x + 4) - (3x + 4)\cos(3x + 4) + 4\cos(3x + 4)) + C$
[Use formula #46 or #48.]

CHAPTER 5: THE INTEGRAL

§5.6 Approximating Sums: The Integral as a Limit

1. (a) Using 5 equal subintervals, the left sum approximation to $\int_{-5}^{5} g(x)\,dx$ is

$$2\big(g(-5) + g(-3) + g(-1) + g(1) + g(3)\big) = 2(-2 + 2 + 3 + 0 + (-1)) = 4;$$

3. (a) Using 5 equal subintervals, the midpoint sum approximation to $\int_{-5}^{5} g(x)\,dx$ is

$$2\big(g(-4) + g(-2) + g(0) + g(2) + g(4)\big) 2(0 + 3 + 2 + (-1) - 0.5) = 7.$$

5. (a) Using 5 equal subintervals, the left sum approximation to $\int_{-1}^{4} g(x)\,dx$ is

$$\big(g(-1) + g(0) + g(1) + g(2) + g(3)\big) = 3 + 2 + 0 - 1 - 1 = 3.$$

7. Using 3 equal subintervals, the right sum approximation to $\int_{1}^{7} f(x)\,dx$ is

$$2\big(f(3) + f(5) + f(7)\big) = 2(3 + 6 + 3) = 24.$$

9. Using 3 equal subintervals, the midpoint sum approximation to $\int_{1}^{7} f(x)\,dx$ is

$$2\big(f(2) + f(4) + f(6)\big) = 2(3 + 6 + 5) = 28.$$

11. Using 4 equal subintervals, the right sum approximation to $\int_{0}^{8} f(x)\,dx$ is

$$2\big(f(2) + f(4) + f(6) + f(8)\big) = 2(3 + 6 + 5 + 0) = 28$$

13. Using 4 equal subintervals, the midpoint sum approximation to $\int_{0}^{8} f(x)\,dx$ is

$$2\big(f(1) + f(3) + f(5) + f(7)\big) = 2(5 + 3 + 6 + 3) = 34.$$

15. $L_4 = \big((-1 + 1) + (0 + 1) + (1 + 1) + (2 + 1)\big) = 6$
    $R_4 = \big((0 + 1) + (1 + 1) + (2 + 1) + (3 + 1)\big) = 10$
    $M_4 = \big((-0.5 + 1) + (0.5 + 1) + (1.5 + 1) + (2.5 + 1)\big) = 8$
    $T_4 = \frac{1}{2}(L_4 + R_4) = 8$

$$I = \int_{-1}^{3} (x + 1)\,dx = \left(\tfrac{1}{2}x^2 + x\right)\Big]_{-1}^{3} = \left(\tfrac{9}{2} + 3\right) - \left(\tfrac{1}{2} - 1\right) = 8$$

17. $L_4 = \big((-2)^3 + (-1)^3 + 0^3 + 1^3\big) = -8$
    $R_4 = \big((-1)^3 + 0^3 + 1^3 + 2^3\big) = 8$
    $M_4 = \big((-1.5)^3 + (-0.5)^3 + (0.5)^3 + (1.5)^3\big) = 0$
    $T_4 = \frac{1}{2}(L_4 + R_4) = 0$

$$I = \int_{-2}^{2} x^3\,dx = \tfrac{1}{4}x^4\Big]_{-2}^{2} = 2^4 - (-2)^4 = 0$$

SECTION 5.6   APPROXIMATING SUMS: THE INTEGRAL AS A LIMIT

19.

n	L_n	R_n	M_n	T_n
2	−0.75000	3.75000	1.5000	1.50000
4	0.37500	2.62500	1.5000	1.50000
8	0.93750	2.06250	1.5000	1.50000
16	1.21875	1.78125	1.5000	1.50000
32	1.35938	1.64063	1.5000	1.50000
64	1.42969	1.57031	1.5000	1.50000
128	1.46484	1.53516	1.5000	1.50000
256	1.48242	1.51758	1.5000	1.50000

$$\int_{-1}^{2} x\, dx = \tfrac{1}{2}x^2 \bigg]_{-1}^{2} = 2 - \tfrac{1}{2} = 3/2 = 1.5$$

21.

n	L_n	R_n	M_n	T_n
2	0.06250	0.56250	0.21875	0.31250
4	0.14063	0.39063	0.24219	0.26563
8	0.19141	0.31641	0.24805	0.25391
16	0.21973	0.28222	0.24951	0.25098
32	0.23462	0.26587	0.24988	0.25024
64	0.24225	0.25787	0.24997	0.25006
128	0.24611	0.25392	0.24999	0.25002
256	0.24805	0.25196	0.25000	0.25000

$$\int_{0}^{1} x^3\, dx = \tfrac{1}{4}x^4 \bigg]_{0}^{1} = \tfrac{1}{4} = 0.25$$

23.

n	L_n	R_n	M_n	T_n
2	1.57080	1.57080	2.22144	1.57080
4	1.89612	1.89612	2.05234	1.89612
8	1.97423	1.97423	2.01291	1.97423
16	1.99357	1.99357	2.00322	1.99357
32	1.99839	1.99839	2.00080	1.99838
64	1.99960	1.99960	2.00020	1.99960
128	1.99990	1.99990	2.00005	1.99990
256	1.99997	1.99997	2.00001	1.99998

$$\int_0^\pi \sin x\, dx = -\cos x \Big]_0^\pi = -\cos \pi + \cos 0 = 2$$

25. $\int_2^5 f(x)\, dx \approx L_3 = f(2) + f(3) + f(4) = 0.21 + 0.28 + 0.36 = 0.85$

27. $\int_1^5 f(x)\, dx \approx R_2 = 2(f(3) + f(5)) = 2(0.28 + 0.44) = 1.44$

29. (a) $\int_5^{20} f(x)\, dx$

    (b) $\int_0^{15} f(x)\, dx$

    (c) $\int_{2.5}^{17.5} f(x)\, dx$

31. The subintervals [2, 4] and [4, 7] partition the interval [2, 7] into subintervals of length 2 and 3, respectively. Therefore, $f(2) \cdot 2 + f(4) \cdot 3$ is a left Riemann sum approximation to $\int_2^7 f(x)\, dx$ with unequal length subintervals.

33. No. The expression $f(3) \cdot 1 + f(4) \cdot 4$ implies that the subintervals of [0, 5] have lengths 1 and 4, respectively. Therefore, the subintervals are [0, 1] and [1, 5]. However, $x = 3$ is not in the first subinterval, so the expression given is not a Riemann sum approximation to $\int_0^5 f(x)\, dx$.

§5.7 Working with Sums

1. $\displaystyle\sum_{k=1}^{50} k$

3. Every odd integer can be written as $(2i + 1)$ for some integer i. Thus,
$$1^2 + 3^2 + 5^2 + \cdots + 97^2 + 99^2 = \sum_{i=0}^{49}(2i+1)^2.$$

5. $\displaystyle\sum_{k=1}^{5} k^{-1} = 1 + \frac{1}{2} + \frac{1}{3} + \frac{1}{4} + \frac{1}{5} = \frac{137}{60}$

7. $\displaystyle\sum_{k=1}^{6} \sin(k\pi/4) = \sin(\pi/4) + \sin(\pi/2) + \sin(3\pi/4) + \sin(\pi) + \sin(5\pi/4) + \sin(3\pi/2)$
$$= \frac{\sqrt{2}}{2} + 1 + \frac{\sqrt{2}}{2} + 0 - \frac{\sqrt{2}}{2} - 1 = \frac{\sqrt{2}}{2}$$

9. $\displaystyle\sum_{k=1}^{17} k = \frac{1}{2}(17)(17+1) = 153$

11. $\displaystyle\sum_{k=0}^{9}(k+1)^2 = \sum_{k=1}^{10}(k+1)^2 = \frac{1}{6}(10)(11)(21) = 385$

13. (a) $f(x) = x^2$, $n = 10$, $a = x_0 = 0$, and $b = x_{10} = 10$, so $\Delta x = (b-a)/10 = 1$. Therefore,
$$\int_0^{10} x^2\,dx \approx R_{10} = \Delta x \sum_{j=1}^{10} f(x_j) = \sum_{j=1}^{10}(x_j)^2 = \sum_{j=1}^{10}(0 + j\Delta x)^2 = \sum_{j=1}^{10} j^2.$$

(b) $f(x) = x^2$, $n = 5$, $a = x_0 = 0$, and $b = x_5 = 10$, so $\Delta x = (b-a)/5 = 2$. Therefore,
$$\int_0^{10} x^2\,dx \approx R_5 = \Delta x \sum_{k=1}^{5} f(x_k) = 2\sum_{k=1}^{5}(x_k)^2 = 2\sum_{k=1}^{5}(0 + k\Delta x)^2 = 2\sum_{k=1}^{5}(2k)^2 = 8\sum_{k=1}^{5} k^2.$$

(c) $f(x) = x^2$, $n = 20$, $a = x_0 = 0$, and $b = x_{20} = 10$, so $\Delta x = (b-a)/20 = 1/2$. Therefore,
$$\int_0^{10} x^2\,dx \approx R_{20} = \Delta x \sum_{i=1}^{20} f(x_i) = \frac{1}{2}\sum_{i=1}^{20}(x_i)^2 = \frac{1}{2}\sum_{i=1}^{20}(0 + i\Delta x)^2 = \frac{1}{2}\sum_{i=1}^{20}\frac{i^2}{4} = \frac{1}{8}\sum_{i=1}^{20} i^2.$$

(d) $f(x) = x^2$, $n = 50$, $a = x_0 = 0$, and $b = x_{50} = 10$, so $\Delta x = (b-a)/50 = 1/5$. Therefore,
$$\int_0^{10} x^2\,dx \approx R_{50} = \Delta x \sum_{m=1}^{50} f(x_m) = \frac{1}{5}\sum_{m=1}^{50}(x_m)^2 = \frac{1}{5}\sum_{m=1}^{50}(0 + m\Delta x)^2 = \frac{1}{5}\sum_{m=1}^{50}\frac{m^2}{25} = \frac{1}{125}\sum_{m=1}^{50} m^2.$$

15. (a) $f(x) = x^2$, $n = 10$, $a = x_0 = 1$, and $b = x_{10} = 11$, so $\Delta x = (b-a)/10 = 1$. Therefore,
$$\int_1^{11} x^2\,dx \approx L_{10} = \Delta x \sum_{i=0}^{9} f(x_i) = \sum_{i=0}^{9}(x_i)^2 = \sum_{i=0}^{9}(1 + i\Delta x)^2 = \sum_{i=0}^{9}(1+i)^2 = \sum_{i=1}^{10} i^2.$$

(b) $f(x) = x^2$, $n = 5$, $a = x_0 = 1$, and $b = x_5 = 11$, so $\Delta x = (b-a)/5 = 2$. Therefore,
$$\int_1^{11} x^2\,dx \approx L_5 = \Delta x \sum_{j=0}^{4} f(x_j) = 2\sum_{j=0}^{4}(x_j)^2 = 2\sum_{j=0}^{4}(1 + j\Delta x)^2 = 2\sum_{j=0}^{4}(1+2j)^2.$$

(c) $f(x) = x^2$, $n = 20$, $a = x_0 = 1$, and $b = x_{20} = 11$, so $\Delta x = (b-a)/20 = 1/2$. Therefore,
$$\int_1^{11} x^2 \, dx \approx L_{20} = \Delta x \sum_{k=0}^{19} f(x_k) = \frac{1}{2} \sum_{k=0}^{19} (x_k)^2 = \frac{1}{2} \sum_{k=0}^{19} (1 + k\Delta x)^2 = \frac{1}{2} \sum_{k=0}^{19} \left(1 + \frac{k}{2}\right)^2.$$

(d) $f(x) = x^2$, $n = 50$, $a = x_0 = 1$, and $b = x_{50} = 11$, so $\Delta x = (b-a)/50 = 1/5$. Therefore,
$$\int_1^{11} x^2 \, dx \approx L_{50} = \Delta x \sum_{m=0}^{49} f(x_m) = \frac{1}{5} \sum_{m=0}^{49} (x_m)^2 = \frac{1}{5} \sum_{m=0}^{49} (1 + m\Delta x)^2 = \frac{1}{5} \sum_{m=0}^{49} \left(1 + \frac{m}{5}\right)^2.$$

17. (a) $f(x) = \sqrt[3]{x} = x^{1/3}$, $n = 10$, $a = x_0 = 0$, and $b = x_{10} = 1$, so $\Delta x = (b-a)/10 = 1/10$. Therefore,
$$\int_0^1 x^{1/3} \, dx \approx L_{10} = \Delta x \sum_{j=0}^{9} f(x_j) = \sum_{j=0}^{9} (x_j)^{1/3} = \sum_{j=0}^{9} (0 + j\Delta x)^{1/3} = \sum_{j=0}^{9} (j/10)^{1/3}.$$

(b) $f(x) = \sqrt[3]{x} = x^{1/3}$, $n = 5$, $a = x_0 = 0$, and $b = x_5 = 1$, so $\Delta x = (b-a)/5 = 1/5$. Therefore,
$$\int_0^1 x^{1/3} \, dx \approx L_5 = \Delta x \sum_{k=0}^{4} f(x_k) = \frac{1}{5} \sum_{k=0}^{4} (x_k)^{1/3} = \frac{1}{5} \sum_{k=0}^{4} (0 + k\Delta x)^{1/3} = \frac{1}{5} \sum_{k=0}^{4} (k/5)^{1/3}.$$

(c) $f(x) = \sqrt[3]{x} = x^{1/3}$, $n = 20$, $a = x_0 = 0$, and $b = x_{20} = 1$, so $\Delta x = (b-a)/20 = 1/20$. Therefore,
$$\int_0^1 x^{1/3} \, dx \approx L_{20} = \Delta x \sum_{i=0}^{19} f(x_i) = \frac{1}{20} \sum_{i=0}^{19} (x_i)^{1/3} = \frac{1}{20} \sum_{i=0}^{19} (0 + i\Delta x)^{1/3} = \frac{1}{20} \sum_{i=0}^{19} (i/20)^{1/3}.$$

(d) $f(x) = \sqrt[3]{x} = x^{1/3}$, $n = 50$, $a = x_0 = 0$, and $b = x_{50} = 1$, so $\Delta x = (b-a)/50 = 1/50$. Therefore,
$$\int_0^1 x^2 \, dx \approx L_{50} = \Delta x \sum_{m=0}^{49} f(x_m) = \frac{1}{50} \sum_{m=0}^{49} (x_m)^{1/3} = \frac{1}{50} \sum_{m=0}^{49} (0 + m\Delta x)^{1/3} = \frac{1}{50} \sum_{m=0}^{49} (m/50)^{1/3}.$$

19. (a) $f(x) = \sqrt[3]{x} = x^{1/3}$, $n = 10$, $a = x_0 = 1$, and $b = x_{10} = 2$, so $\Delta x = (b-a)/10 = 1/10$. Therefore,
$$\int_1^2 x^{1/3} \, dx \approx R_{10} = \Delta x \sum_{j=1}^{10} f(x_j) = \frac{1}{10} \sum_{j=1}^{10} (x_j)^{1/3} = \frac{1}{10} \sum_{j=1}^{10} (1 + j\Delta x)^{1/3} = \frac{1}{10} \sum_{j=1}^{10} (1 + j/10)^{1/3}.$$

(b) $f(x) = \sqrt[3]{x} = x^{1/3}$, $n = 5$, $a = x_0 = 1$, and $b = x_5 = 2$, so $\Delta x = (b-a)/5 = 1/5$. Therefore,
$$\int_1^2 x^{1/3} \, dx \approx R_5 = \Delta x \sum_{k=1}^{5} f(x_k) = \frac{1}{5} \sum_{k=1}^{5} (x_k)^{1/3} = \frac{1}{5} \sum_{k=1}^{5} (1 + k\Delta x)^{1/3} = \frac{1}{5} \sum_{k=1}^{5} (1 + k/5)^{1/3}.$$

(c) $f(x) = \sqrt[3]{x} = x^{1/3}$, $n = 20$, $a = x_0 = 1$, and $b = x_{20} = 2$, so $\Delta x = (b-a)/20 = 1/20$. Therefore,
$$\int_1^2 x^{1/3} \, dx \approx R_{20} = \Delta x \sum_{i=1}^{20} f(x_i) = \frac{1}{20} \sum_{i=1}^{20} (x_i)^{1/3} = \frac{1}{20} \sum_{i=1}^{20} (1 + i\Delta x)^{1/3} = \frac{1}{20} \sum_{i=1}^{20} (1 + i/20)^{1/3}.$$

(d) $f(x) = \sqrt[3]{x} = x^{1/3}$, $n = 50$, $a = x_0 = 1$, and $b = x_{50} = 2$, so $\Delta x = (b-a)/50 = 1/50$. Therefore,
$$\int_1^2 x^{1/3} \, dx \approx R_{50} = \Delta x \sum_{m=1}^{50} f(x_m) = \frac{1}{50} \sum_{m=1}^{50} (x_m)^{1/3} = \frac{1}{50} \sum_{m=1}^{50} (1 + m\Delta x)^{1/3} = \frac{1}{50} \sum_{m=1}^{50} (1 + m/50)^{1/3}.$$

21. (a) $f(x) = \sin(x^2)$, $n = 10$, $a = x_0 = 3$, and $b = x_{10} = 7$, so $\Delta x = (b-a)/10 = 2/5$. Therefore,
$$\int_3^7 \sin(x^2) \, dx \approx R_{10} = \Delta x \sum_{j=1}^{10} f(x_j) = \frac{2}{5} \sum_{j=1}^{10} \sin(x_j)^2 = \frac{2}{5} \sum_{j=1}^{10} \sin(3 + j\Delta x)^2 = \frac{2}{5} \sum_{j=1}^{10} \sin(3 + 2j/5)^2.$$

(b) $f(x) = \sin(x^2)$, $n = 5$, $a = x_0 = 3$, and $b = x_5 = 7$, so $\Delta x = (b-a)/5 = 4/5$. Therefore,
$$\int_3^7 \sin(x^2) \, dx \approx R_5 = \Delta x \sum_{k=1}^{5} f(x_k) = \frac{4}{5} \sum_{k=1}^{5} \sin(x_k)^2 = \frac{4}{5} \sum_{k=1}^{5} \sin(3 + k\Delta x)^2 = \frac{4}{5} \sum_{k=1}^{5} \sin(3 + 4k/5)^2.$$

(c) $f(x) = \sin(x^2)$, $n = 20$, $a = x_0 = 3$, and $b = x_{20} = 7$, so $\Delta x = (b-a)/20 = 1/5$. Therefore,
$$\int_3^7 \sin(x^2) \, dx \approx R_{20} = \Delta x \sum_{i=1}^{20} f(x_i) = \frac{1}{5} \sum_{i=1}^{20} \sin(x_i)^2 = \frac{1}{5} \sum_{i=1}^{20} \sin(3 + i\Delta x)^2 = \frac{1}{5} \sum_{i=1}^{20} \sin(3 + i/5)^2.$$

Section 5.7 Working with Sums

(d) $f(x) = \sin(x^2)$, $n = 50$, $a = x_0 = 3$, and $b = x_{50} = 7$, so $\Delta x = (b-a)/50 = 2/25$. Therefore,

$$\int_3^7 \sin(x^2)\,dx \approx R_{50} = \Delta x \sum_{m=1}^{50} f(x_m) = \frac{2}{25}\sum_{m=1}^{50}(x_m)^2 = \frac{2}{25}\sum_{m=1}^{50}\sin(3+m\Delta x)^2 = \frac{2}{25}\sum_{m=1}^{50}\sin(3+\tfrac{2m}{25})^2.$$

23. (a) $f(x) = \sqrt{x} = x^{1/2}$, $n = 10$, $a = x_0 = 0$, and $b = x_{10} = 5$, so $\Delta x = (b-a)/10 = 1/2$. Therefore,

$$\int_0^5 x^{1/2}\,dx \approx M_{10} = \Delta x\sum_{j=0}^{9} f\big((x_j+x_{j+1})/2\big) = \frac{1}{2}\sum_{j=0}^{9}(1/4+j\Delta x)^{1/2} = \frac{1}{2}\sum_{j=0}^{9}(1/4+j/2)^{1/2}.$$

(b) $f(x) = \sqrt{x} = x^{1/2}$, $n = 5$, $a = x_0 = 0$, and $b = x_5 = 5$, so $\Delta x = (b-a)/5 = 1$. Therefore,

$$\int_0^5 x^{1/2}\,dx \approx M_5 = \Delta x\sum_{k=0}^{4} f\big((x_k+x_{k+1})/2\big) = \sum_{k=0}^{4}(1/2+k\Delta x)^{1/2} = \sum_{k=0}^{4}(1/2+k)^{1/2}.$$

(c) $f(x) = \sqrt{x} = x^{1/2}$, $n = 20$, $a = x_0 = 0$, and $b = x_{20} = 5$, so $\Delta x = (b-a)/20 = 1/4$. Therefore,

$$\int_0^5 x^{1/2}\,dx \approx M_{20} = \Delta x\sum_{i=0}^{19} f\big((x_i+x_{i+2})/2\big) = \frac{1}{4}\sum_{i=0}^{19}(1/8+i\Delta x)^{1/2} = \frac{1}{4}\sum_{i=0}^{19}(1/8+i/4)^{1/2}.$$

(d) $f(x) = \sqrt{x} = x^{1/2}$, $n = 50$, $a = x_0 = 0$, and $b = x_{50} = 5$, so $\Delta x = (b-a)/50 = 1/10$. Therefore,

$$\int_0^5 x^{1/2}\,dx \approx M_{50} = \Delta x\sum_{m=0}^{49} f\big((x_m+x_{m+1})/2\big) = \frac{1}{10}\sum_{m=0}^{49}(\tfrac{1}{20}+m\Delta x)^{1/2} = \frac{1}{10}\sum_{m=0}^{49}(\tfrac{1}{20}+\tfrac{m}{10})^{1/2}.$$

25. Since $\sum_{j=1}^{n} j = \tfrac{1}{2}n(n+1)$,

$$\lim_{n\to\infty}\frac{1+2+3+\cdots+n}{n^2} = \lim_{n\to\infty}\frac{n(n+1)}{2n^2} = \lim_{n\to\infty}\frac{n^2+n}{2n^2} = \lim_{n\to\infty}\left(\frac{1}{2}+\frac{1}{2n}\right) = \frac{1}{2}.$$

27. The right sum approximation to $\int_4^5 x^3\,dx$ with n equal subintervals is $R_n = \Delta x\sum_{k=1}^{n} f(x_k)$, where

$\Delta x = (5-4)/n = 1/n$, $f(x) = x^3$, and $x_k = 4+k\Delta x = 4+k/n$, so $R_n = \dfrac{1}{n}\sum_{k=1}^{n}(4+k/n)^3$. Therefore,

by the definition of the integral as a limit of Riemann sums, $\lim_{n\to\infty} R_n = \int_4^5 x^3\,dx$.

29. The right sum approximation to $\int_1^4 \sqrt{x}\,dx$ with n equal subintervals is $R_n = \Delta x\sum_{k=1}^{n} f(x_k)$, where

$\Delta x = (4-1)/n = 3/n$, $f(x) = \sqrt{x}$, and $x_k = 1+k\Delta x = 1+3k/n$, so $R_n = \dfrac{3}{n}\sum_{k=1}^{n}\sqrt{1+3k/n}$.

Therefore, by the definition of the integral as a limit of Riemann sums, $\lim_{n\to\infty} R_n = \int_1^4 \sqrt{x}\,dx$.

31. The left sum approximation to $\int_1^4 \sqrt{x}\,dx$ with n equal subintervals is $L_n = \Delta x\sum_{k=0}^{n-1} f(x_k)$, where

$\Delta x = (4-1)/n = 3/n$, $f(x) = \sqrt{x}$, and $x_k = 1+k\Delta x = 1+3k/n$, so

$L_n = \dfrac{3}{n}\sum_{k=0}^{n-1}\sqrt{1+3k/n} = \dfrac{3}{n}\sum_{k=1}^{n}\sqrt{1+3(k-1)/n}$. Therefore, by the definition of the integral as a limit of

Riemann sums, $\lim_{n\to\infty} L_n = \int_1^4 \sqrt{x}\,dx$.

CHAPTER 5: THE INTEGRAL

33. The midpoint sum approximation to $\int_1^4 \sqrt{x}\, dx$ with n equal subintervals is

$$M_n = \Delta x \sum_{k=0}^{n-1} f\big((x_k + x_{k+1})/2\big) = M_n = \Delta x \sum_{k=1}^{n} f\big((x_{k-1} + x_k)/2\big), \text{ where } \Delta x = (4-1)/n = 3/n,$$

$f(x) = \sqrt{x}$, and $x_k = 1 + k\Delta x = 1 + 3k/n$, so

$$M_n = \frac{3}{n} \sum_{k=1}^{n} \sqrt{\big(1 + 3(k-1)/2n\big) + \big(1 + 3k/2n\big)} = \frac{3}{n} \sum_{k=1}^{n} \sqrt{1 + 3(2k-1)/2n}. \text{ Therefore, by the}$$

definition of the integral as a limit of Riemann sums, $\lim_{n\to\infty} M_n = \int_1^4 \sqrt{x}\, dx$.

35. $\int_2^5 x\, dx = \lim_{n\to\infty} \frac{3}{n} \sum_{j=0}^{n-1} (2 + 3j/n) = \lim_{n\to\infty} \frac{3}{n} \left(2n + \frac{3}{n} \cdot \frac{(n-1)(n)}{2}\right) = 6 + \frac{9}{2} = \frac{21}{2}$

37. $\int_0^1 x^2\, dx = \lim_{n\to\infty} \frac{1}{n} \sum_{k=0}^{n-1} \left(\frac{k}{n}\right)^2 = \lim_{n\to\infty} \frac{1}{n^3} \sum_{k=0}^{n-1} k^2 = \lim_{n\to\infty} \frac{1}{n^3} \cdot \frac{1}{6}(n-1)(n)(2(n-1)+1)$

$= \lim_{n\to\infty} \frac{2n^3 - 3n^2 + n}{6n^3} = \frac{1}{3}$

39. $\frac{2}{100} \sum_{k=1}^{100} \sin\left(\frac{2k}{100}\right) \approx \int_0^2 \sin x\, dx = -\cos 2 + \cos 0 \approx 1.41615$. (This is a right sum approximation to the integral.)

41. $\lim_{n\to\infty} \frac{1}{n} \sum_{j=1}^{n} \left(\frac{j}{n}\right)^3 = \int_0^1 x^3\, dx = \frac{1}{4}$

43. The average value of x^3 over the interval $[2, 7]$ is $\left(\int_2^7 x^3\, dx\right)/5$. Now, the right sum approximation to $\int_2^7 x^3\, dx$ is $R_n = \frac{5}{n} \sum_{k=1}^{n} (2 + 5k/n)^3$. Therefore, the average value of x^3 over the interval $[2, 7]$ is

$$\frac{1}{5} \int_2^7 x^3\, dx = \frac{1}{5} \cdot \lim_{n\to\infty} \frac{5}{n} \sum_{k=1}^{n} (2 + 5k/n)^3 = \lim_{n\to\infty} \frac{1}{n} \sum_{k=1}^{n} (2 + 5k/n)^3.$$

45. (b) To estimate the total distance traveled:

   (i) A trapezoid approximating sum, 6 subdivisions:

   $$T_6 = \frac{1}{6}(42/2 + 38 + 36 + 57 + 0 + 55 + 51/2) = 38.75$$

   (ii) A left approximating sum, 6 subdivisions:

   $$L_6 = \frac{1}{6}(42 + 38 + 36 + 57 + 0 + 55) = 38$$

   (iii) A midpoint approximating sum, 3 subdivisions

   $$M_6 = \frac{1}{3}(38 + 57 + 55) = 50$$

   The trapezoid rule answer may be best, since that estimate also approximates the change in speed during each of the 10-minute intervals.

SECTION 5.7 WORKING WITH SUMS

(c) To plot a plausible distance graph, one might estimate the distances covered over each 10-minute period (using trapezoids) and add them up:

Distance estimates over one hour							
time (min)	0	10	20	30	40	50	60
speed (mph)	42	38	36	57	0	55	51
total distance (miles)	0	6.66	12.83	20.58	25.33	29.92	38.75

47. (a) $\sum_{k=1}^{n}\left((k+1)^2 - k^2\right) = (2^2 - 1^2) + (3^2 - 2^2) + \cdots + \left(n^2 - (n-1)^2\right) + \left((n+1)^2 - n^2\right)$

    $= (n+1)^2 - 1$

(b) $(k+1)^2 - k^2 = (k^2 + 2k + 1) - k^2 = 2k + 1$

(c) $\sum_{k=1}^{n}(2k+1) = 2\sum_{k=1}^{n}k + \sum_{k=1}^{n}1 = 2\sum_{k=1}^{n}k + n$ and $(n+1)^2 - 1 = n^2 + 2n$ so

$\sum_{k=1}^{n}(2k+1) = (n+1)^2 - 1 \implies 2\sum_{k=1}^{n}k = n^2 + n = n(n+1) \implies \sum_{k=1}^{n}k = \frac{1}{2}n(n+1)$. Since

$\sum_{k=1}^{n}(2k+1) = (n+1)^2 - 1,$

49. $\sum_{k=1}^{n}\left((k+1)^4 - k^4\right) = (n+1)^4 - 1$ and $(k+1)^4 - k^4 = 4k^3 + 6k^2 + 4k + 1$. Therefore,

$(n+1)^4 - 1 = \sum_{k=1}^{n}\left((k+1)^4 - k^4\right) = \sum_{k=1}^{n}(4k^3 + 6k^2 + 4k + 1)$

$= 4\sum_{k=1}^{n}k^3 + 6\sum_{k=1}^{n}k^2 + 4\sum_{k=1}^{n}k + \sum_{k=1}^{n}1$

$= 4\sum_{k=1}^{n}k^3 + (2n^3 + 3n^2 + n) + (2n^2 + 2n) + n = 4\sum_{k=1}^{n}k^3 + 2n^3 + 5n^2 + 4n$

$\implies \sum_{k=1}^{n}k^3 = \frac{1}{4}\left((n+1)^4 - 2n^3 - 5n^2 - 4n - 1\right) = \frac{1}{4}(n^4 + 2n^3 + n^2)$

$= \frac{1}{4}n^2(n+1)^2.$

§5.8 Chapter Summary

1. (a) Note that $f(x) < 0$ for all x in the interval $[0, 40]$. Furthermore, it is clear from the picture that $f(x) < -55$ (the area of eleven dotted rectangles). Thus, the best estimate of $\int_0^{40} f(x)\, dx$ is -65.

   (b) $\dfrac{\int_{10}^{30} f(x)\, dx}{30 - 10} \approx \dfrac{-37}{20} = -1.85$

3. (a) $y = -2$ is one possibility.

   (b) $y = 8 - 20x$ is one possibility.

5. If $0 \le x \le \pi$, $\cos 1 \le \cos(\sin x) \le 1$. Therefore,
$$\frac{\pi}{2} < \cos 1 \cdot (\pi - 0) \le \int_0^\pi \cos(\sin x)\, dx \le 1 \cdot (\pi - 0) = \pi.$$

7. (a) Since the graph of the arctangent function is concave down, the secant line through the points $(0, \arctan 0) = (0, 0)$ and $(1, \arctan 1) = (1, \pi/4)$ lies below the curve $y = \arctan x$.

   (b) Let $f(x) = \arctan x$. Then $f(0) = 0$, $f'(x) = 1/(1 + x^2)$, and so $f'(0) = 1$. Therefore, the equation of the tangent line is $y = x$.

   (c) Let g be secant line from part (a) [$g(x) = \pi x/4$] and $h(x) = x$ be the tangent line from part (b). Then $g(x) \le \arctan x \le h(x)$ if $0 \le x \le 1$. Therefore,
$$\frac{\pi}{8} = \int_0^1 g(x)\, dx \le \int_0^1 \arctan x\, dx \le \int_0^1 h(x)\, dx = \frac{1}{2}.$$

9. (a) $\int_0^{\pi/2} \sin^2 x\, dx + \int_0^{\pi/2} \cos^2 x\, dx = \int_0^{\pi/2} \left(\sin^2 x + \cos^2 x\right) dx = \int_0^{\pi/2} dx = \dfrac{\pi}{2}$

   (b) $\int_0^{\pi/2} \cos^2 x\, dx = \int_0^{\pi/2} \sin^2(x - \pi/2)\, dx = \int_{-\pi/2}^0 \sin^2 x\, dx = \int_0^{\pi/2} \sin^2 x\, dx$

   (c) $\int_0^{\pi/2} \sin^2 x\, dx + \int_0^{\pi/2} \cos^2 x\, dx = 2 \int_0^{\pi/2} \sin^2 x\, dx = \dfrac{\pi}{2}$, so $\int_0^{\pi/2} \sin^2 x\, dx = \dfrac{\pi}{4}$.

11. The bounds on f imply that $-4 \le \int_1^3 f(x)\, dx \le 10$. Thus, $-2 \le \dfrac{\int_1^3 f(x)\, dx}{2} \le 5$.

13. (a) No. Let $f(x) = 0$ and $g(x) = x$. Then $\int_{-1}^1 f(x)\, dx = \int_{-1}^1 g(x)\, dx = 0$ but $f(x) \ge g(x)$ if $-1 \le x \le 0$.

    (b) Yes. If $f(x) > g(x)$ for *every* x such that $a \le x \le b$, then $\int_a^b f(x)\, dx > \int_a^b g(x)\, dx$ would be true — a contradiction.

15. $\sin^2 x = 1 - \cos^2 x \implies \sin^2 x + C_1 = (1 - \cos^2 x) + C_1 = -\cos^2 x + (1 + C_1) = -\cos^2 x + C_2$

17. $\dfrac{1}{2} \cos x \sin x = \dfrac{2 \cos x \sin x}{4} = \dfrac{1}{4} \sin(2x)$

19. $\left(x \ln x - x + C\right)' = \ln x + \dfrac{x}{x} - 1 = \ln x$

21. $\left(\dfrac{1}{a} \arctan\left(\dfrac{x}{a}\right) + C\right)' = \dfrac{1}{a} \cdot \dfrac{1/a}{1 + (x/a)^2} = \dfrac{1}{a^2 + x^2}$

23. $\left(\ln|\sec x + \tan x| + C\right)' = \dfrac{\sec x \tan x + \sec^2 x}{\sec x + \tan x} = \dfrac{\sec x (\tan x + \sec x)}{\sec x + \tan x} = \sec x$

Section 5.8 Chapter Summary

25. $\left(x \arcsin x + \sqrt{1-x^2} + C\right)' = \arcsin x + \dfrac{x}{\sqrt{1-x^2}} + \dfrac{-2x}{2\sqrt{1-x^2}} = \arcsin x$

27. $\displaystyle\int \dfrac{dx}{3x} = \tfrac{1}{3} \ln |x| + C$

29. $\displaystyle\int \dfrac{3}{x^2+1} dx = 3 \arctan x + C$

31. $\displaystyle\int (2 \sin(3x) - 4 \cos(5x)) dx = -\tfrac{2}{3} \cos(3x) - \tfrac{4}{5} \sin(5x) + C$

33. $\displaystyle\int \left(1+\sqrt{x}\right)^2 dx = \int (1 + 2\sqrt{x} + x) dx = x + \tfrac{4}{3} x^{3/2} + \tfrac{1}{2} x^2 + C$

35. $\displaystyle\int \dfrac{(3-x)^2}{x} dx = \int \dfrac{9 - 6x + x^2}{x} dx = 9 \ln |x| - 6x + \tfrac{1}{2} x^2 + C$

37. $\displaystyle\int e^x (1 - e^x) dx = \int \left(e^x - e^{2x}\right) = e^x - \tfrac{1}{2} e^{2x} + C$

39. Let $u = x^2$. Then $du = 2x\, dx$ so $\displaystyle\int x \sin(x^2) dx = \dfrac{1}{2} \int \sin u\, du = -\tfrac{1}{2} \cos u + C = -\tfrac{1}{2} \cos(x^2) + C$.

41. Let $u = x^4$. Then $du = 4x^3\, dx$ so $\displaystyle\int x^3 e^{x^4} dx = \dfrac{1}{4} \int e^u\, du = \tfrac{1}{4} e^u + C = \tfrac{1}{4} e^{x^4} + C$.

43. Using the substitution $u = \ln x$, $\displaystyle\int \dfrac{(\ln x)^3}{x} dx = \int u^3\, du = \dfrac{1}{4} u^4 = \dfrac{1}{4} (\ln x)^4 + C$.

45. Let $u = \ln(\cos x)$. Then $du = -\dfrac{\sin x}{\cos x} dx = -\tan x\, dx$. Therefore, $\displaystyle\int \ln(\cos x) \tan x\, dx = -\int u\, du$
    $= -\tfrac{1}{2} u^2 + C = -\tfrac{1}{2} \left(\ln |\cos x|\right)^2 + C$.

47. Let $u = 1 - x$. Then, $\displaystyle\int_0^1 x^n (1-x)^m dx = \int_1^0 (1-u)^n u^m (-du) = \int_0^1 u^m (1-u)^n du$.

49. $\sqrt{1 - \sin^2 x} = -\cos x$ if $\pi/2 \le x \le \pi$. The "proof" incorrectly replaces $\sqrt{1 - \sin^2 x}$ by $\cos x$ for all x in the interval $[0, \pi]$.

51. According to the Fundamental Theorem of Calculus, the change in the coffee's temperature is

$$\Delta T = \int_0^5 T'(t)\, dt = \int_0^5 \left(-3.5 e^{-0.05 t}\right) dt = \left.\dfrac{3.5}{0.05} e^{-0.05 t}\right|_0^5 = 70 \left(e^{-0.25} - 1\right) \approx -15.5 \,°C.$$

Therefore, after 5 minutes, the temperature of the coffee is approximately $74.5\,°C$.

53. left: $\displaystyle\int_0^5 \sqrt{3x}\, dx \approx \dfrac{5}{N} \sum_{k=1}^N \sqrt{3 \cdot (k-1) \cdot \dfrac{5}{N}} = \dfrac{5}{N} \sum_{j=0}^{N-1} \sqrt{3 \cdot j \cdot \dfrac{5}{N}}$

right: $\displaystyle\int_0^5 \sqrt{3x}\, dx \approx \dfrac{5}{N} \sum_{k=1}^N \sqrt{3 \cdot k \cdot \dfrac{5}{N}} = \dfrac{5}{N} \sum_{j=0}^{N-1} \sqrt{3 \cdot (j+1) \cdot \dfrac{5}{N}}$

midpoint: $\displaystyle\int_0^5 \sqrt{3x}\, dx \approx \dfrac{5}{N} \sum_{k=1}^N \sqrt{3 \cdot (k - 0.5) \cdot \dfrac{5}{N}} = \dfrac{5}{N} \sum_{j=0}^{N-1} \sqrt{3 \cdot (j + 0.5) \cdot \dfrac{5}{N}}$

55. $\int_1^3 g(u)\,du = G(3)$

57. $\int_{-2}^2 g(t)\,dt = \int_{-2}^1 g(t)\,dt + \int_1^2 g(t)\,dt = -\int_1^{-2} g(t)\,dt + \int_1^2 g(t)\,dt$
$= -G(-2) + G(2) = G(2) - G(-2)$

59. (a) If f is an *even* function, $f(x) = f(-x)$ so the signed area of the region enclosed by the graph of f between $x = -3$ and $x = 0$ is the same as the signed area of the region enclosed by the graph of f between $x = 0$ and $x = 3$. Thus,
$\int_{-3}^3 f(x)\,dx = \int_{-3}^0 f(x)\,dx + \int_0^3 f(x)\,dx = 2\int_0^3 f(x)\,dx = 2 \cdot -1 = -2$.

(b) If f is an *odd* function, $f(x) = -f(-x)$ so the signed area of the region enclosed by the graph of f between $x = -3$ and $x = 0$ has the same magnitude but the opposite sign as the signed area of the region enclosed by the graph of f between $x = 0$ and $x = 3$. Thus,
$\int_{-3}^3 f(x)\,dx = \int_{-3}^0 f(x)\,dx + \int_0^3 f(x)\,dx = 0$.

61. Note that $h(1) = \int_0^1 g(t)\,dt$, that $h(4) = \int_0^4 g(t)\,dt$, that $\int_1^1 g(x)\,dx = 0$, and that
$\int_5^3 g(t)\,dt = -\int_3^5 g(t)\,dt$. Finally, using the signed area interpretation of the definite integral, it is clear from the graph that $\int_5^3 g(t)\,dt < \int_0^2 g(u)\,du < -1 < h(1) < \int_1^1 g(x)\,dx < h(4) < 3 < 5$.

63. No — Since $h(x) = \int_0^x g(t)\,dt$, $h'(x) = g(x)$ and $h''(x) = g'(x)$. Therefore, $h''(3) = g'(3) > 0$ since g is increasing at $x = 3$.

65. $\int_4^6 (2T(z) + 3)\,dz = 2\int_4^6 T(z)\,dz + 3\int_4^6 dz = 2 \cdot 15 + 3 \cdot (6 - 4) = 36$

67. $\int_1^{-3} h(w)\,dw = -\int_{-3}^1 h(w)\,dw = -(-2) = 2$

69. $\int_{-3}^{-1} h(z)\,dz = \int_{-3}^1 h(t)\,dt - \int_{-1}^1 h(u)\,du = -2 - 4 = -6$

71. (a) $H(-3) = -\int_{-3}^{-2} xe^x\,dx > 0$

[NOTE: Since $-3 \le x \le -2 \implies xe^x < 0$, $\int_{-3}^{-2} xe^x\,dx < 0$.]

(b) $H(-2) = \int_{-2}^{-2} xe^x\,dx = 0$

(c) $H(0) = \int_{-2}^0 xe^x\,dx < 0$
[NOTE: $xe^x < 0$ for all $x < 0$.]

(d) $H(2) = \int_{-2}^2 xe^x\,dx > 0$
[NOTE: A quick look at the graph of $h(x) = xe^x$ over the interval $[-2, 2]$ makes it clear that
$\left|\int_{-2}^0 xe^x\,dx\right| < \int_0^2 xe^x\,dx$. Therefore, $\int_{-2}^2 xe^x\,dx = \int_{-2}^0 xe^x\,dx + \int_0^2 xe^x\,dx > 0$.]

73. $H'(w) = we^w$, so $H'(1) = e$.

SECTION 5.8 CHAPTER SUMMARY

75. $\left(\dfrac{3}{n}\right) \sum_{k=1}^{n} \left(1 + \dfrac{3k}{n}\right)^2$ is a right Riemann sum approximation to the integral $\int_1^4 x^2\, dx$. Therefore,

$$\lim_{n \to \infty} \left(\dfrac{3}{n}\right) \sum_{k=1}^{n} \left(1 + \dfrac{3k}{n}\right)^2 = \int_1^4 x^2\, dx = \dfrac{x^3}{3}\bigg]_1^4 = 21.$$

77. Since $G(x) = \int_x^2 \cos(\pi t^2/4)\, dt = -\int_2^x \cos(\pi t^2/4)\, dt$, $G'(x) = -\cos(\pi x^2/4)$. Therefore, $G'(2) = -\cos(\pi) = 1$.

79. Yes — it is L_7, the left Riemann sum approximation of the integral $\int_3^5 \sqrt{x}\, dx$ computed using $n = 7$ equal subintervals (i.e., $\Delta x = (5-3)/7 = 2/7$):

$$\dfrac{2}{7} \sum_{k=0}^{6} \sqrt{3 + 2k/7} = \dfrac{2}{7}\Big(\sqrt{3} + \sqrt{3 + 2/7} + \sqrt{3 + 4/7} + \sqrt{3 + 6/7}$$
$$+ \sqrt{3 + 8/7} + \sqrt{3 + 10/7} + \sqrt{3 + 12/7}\Big).$$

81. $\int_4^1 (2h(z) - 5)\, dz = -\int_1^4 (2h(z) - 5)\, dz = -2\int_1^4 h(z)\, dz + 5\int_1^4 dz = -34 + 15 = -19$

83. By examining a graph of $\sin(t^2/2)$ over the interval $[0, 3]$ and using the signed area interpretation of the definite integral, one can see that $-0.5 < S(0) < S(1) < 0.5 < S(3) < S(2)$.

85. $S'(z) = \sin(z^2/2)$ so $S''(z) = z\cos(z^2/2)$. Therefore, $S''(3) = 3\cos(9/2) \approx -0.632 < 0$. This implies that $S(z)$ is concave down at $z = 3$.

87. Using the signed area interpretation of the integral and a graph of the integrand, it is clear that $S(z) = 0$ for only one value of z in the interval $[0, 5]$ (i.e., at $z = 1$).

89. The maximum value of S over the interval must occur at a stationary point of S or at an endpoint of the interval. Now, $S'(z) = \sin(z^2/2)$ changes sign from positive to negative only at $z = \sqrt{2\pi}$ and at $z = \sqrt{6\pi}$, so these are the only points in the interior of the interval $[0, 5]$ at which the maximum value could occur. Using the signed area interpretation of the definite integral, it is clear from a graph of the integrand that $S(0) < 0 < S(5) < S(\sqrt{6\pi}) < S(\sqrt{2\pi})$. Thus, S attains its maximum value over the interval $[0, 5]$ at $z = \sqrt{2\pi}$.

91. $\int_2^1 f(t)\, dt = -\int_1^2 f(t)\, dt = -3$

93. insufficient information is given

95. $\int_2^3 f(t)\, dt = \int_1^3 f(t)\, dt - \int_1^2 f(t)\, dt = 7 - 3 = 4$

97. The average value of the function g over the interval $[-2, 3]$ is

$$\dfrac{\int_{-2}^3 g(z)\, dz}{3 - (-2)} = \dfrac{1}{5}\left(\int_{-2}^{-1}(-1-z)\, dz + \int_{-1}^{3}(1+z)\, dz\right) = \dfrac{1}{5}\left(\dfrac{1}{2} + 8\right) = 1.7.$$

99. $\dfrac{3}{n} \sum_{j=1}^{n} \cos\left(1 + \dfrac{3j}{n}\right)$ is a right Riemann sum approximation to $\int_1^4 \cos x\, dx$. Therefore,

$$\lim_{n \to \infty} \dfrac{3}{n} \sum_{j=1}^{n} \cos\left(1 + \dfrac{3j}{n}\right) = \int_1^4 \cos x\, dx = \sin x\bigg]_1^4 = \sin 4 - \sin 1.$$

101. $\int_7^9 f'(x)\,dx = f(9) - f(7)$. Since f is increasing on the interval $(2, \infty)$, $f(9) > f(7)$ so $f(9) - f(7) > 0$.

103. $F''(3) = f'(3) > 0$ since f is increasing on the interval $(2, \infty)$. Therefore, $F(x)$ is concave up at $x = 3$.

105. $\int_0^1 \dfrac{dx}{\sqrt{3 - 2x - x^2}} = \int_0^1 \dfrac{dx}{\sqrt{4 - (x+1)^2}} = 2\int_{1/2}^1 \dfrac{du}{\sqrt{4 - 4u^2}} = \int_{1/2}^1 \dfrac{du}{\sqrt{1 - u^2}}$
$= \arcsin 1 - \arcsin(1/2) = \dfrac{\pi}{2} - \dfrac{\pi}{6} = \dfrac{\pi}{3}$

107. $\dfrac{d}{dx}\left(\int_0^x f(t)\,dt\right) = f(x) \neq \int_0^x f'(t)\,dt = f(x) - f(0)$ since $f(0) \neq 0$.

109. $\int_0^x h(t)\,dt = x^2 - e^x$ implies that $\int_0^0 h(t)\,dt = -1$, but $\int_0^0 h(t)\,dt = 0$ must be true.

111. Yes. Let $x_0 = 0$, $x_1 = 1$, $x_2 = 3$, $x_3 = 6$, $x_4 = 10$, $c_1 = 0$, $c_2 = 2$, $c_3 = 4$, and $c_4 = 6$. Since the x_k's partition the interval $[0, 10]$ and $x_0 \leq c_1 \leq x_1 \leq c_2 \leq x_2 \leq c_3 \leq x_3 \leq c_4 \leq x_4$, $\displaystyle\sum_{k=1}^{4} f(c_k)(x_k - x_{k-1})$ *is a* Riemann sum approximation to $\int_0^{10} f(x)\,dx$.

113. No. $g(3) = \int_0^3 \dfrac{e^{-t}}{1 + t^2}\,dt < \int_0^3 e^{-t}\,dt = 1 - e^{-3} < 1$. Alternatively,
$g(3) < \int_0^3 \dfrac{dt}{1 + t^2} = \arctan 3 < 1.25$.

115. Yes. Since the integrand is positive over the interval $[0, \infty)$, g is increasing over this interval. Althernatively, $g'(x) = \dfrac{e^{-x}}{1 + x^2}$, so $g'(1) = e^{-1}/5 > 0$ which implies that $g(x)$ is increasing at $x = 1$.

117. Let $u = x^3$. Then $du = 3x^2\,dx$ so $\int_0^2 x^2 f(x^3)\,dx = \dfrac{1}{3}\int_0^8 f(u)\,du = \dfrac{4}{3}$.

119. Using the geometric formula for the area of a triangle (twice),
$$\int_{-1}^2 |z|\,dz = \dfrac{1}{2}\cdot 1 + 1\cdot 2 = \dfrac{5}{2}.$$
Alternatively,
$$\int_{-1}^2 |z|\,dz = \int_{-1}^0 (-z)\,dz + \int_0^2 z\,dz = \dfrac{1}{2} + 2 = \dfrac{5}{2}.$$

APPENDIX A: MACHINE GRAPHICS

Appendix A

1. (a)

x	$\sin x$	$\|x - \sin x\|$	x	$\sin x$	$\|x - \sin x\|$
-1	-0.84	0.16	1	0.84	0.16
-0.5	-0.48	0.02	0.5	0.48	0.02
-0.1	-0.10	0	0.10	0.10	0
-0.01	-0.01	0	0.01	0.01	0

(b) The number 0.16 can replace 0.009. (So, of course, can any larger number.)

(c) This can be done by graphing the expression $|p(x) - \sin x|$ over the interval $[-0.65, 0.65]$.

(d) $(-a, a) = (-0.18, 0.18)$—i.e., $a = 0.18$—will work here. Any smaller value of a will also do the trick.

3. It's important to notice that although the picture looks the same as in the previous exercise, the units on the x-axis have changed. The effect is that slopes are *divided* by 10.

   (a) No lines have slope greater than 1.

   (b) Line **A** is described by the equation $y = 2x/15 - 1/2$ and line **D** is described by the equation $y = -x/40 - 1/8$. These lines intersect at the point $(45/19, -7/38)$.

5. Each answer below is only one possibility—there are others.

   (a) Choosing *xrange* $[1.56, 1.58]$ and *yrange* $[0, 2]$ is one possibility. Thus the window $[1.56, 1.58] \times [0, 2]$ works.

   (b) $[-0.1, 0.1] \times [-0.1, 0.1]$

   (c) $[3.1, 3.2] \times [-0.05, 0.05]$

   (d) $[1, 5] \times [-0.01, 0.01]$

   (e) $[-\pi, 0] \times [-1, 0]$

7. Each answer below is only one possibility—there are others.

   (a) Choosing *xrange* $[-0.1, 0.1]$ and *yrange* $[-1, 1]$ is one possibility. Thus the window $[-0.1, 0.1] \times [-1, 1]$ works.

   (b) $[0.9, 1.1] \times [0.8, 1.2]$

   (c) $[-1.1, -0.9] \times [0.8, 1.2]$

   (d) $[0, 10] \times [24, 26]$

   (e) $[-1, 1] \times [0, 1]$

9. The roots are $x \approx -0.767$, $x = 2$ and $x = 4$. (The last two are exact.)

11. The only root is $x \approx 0.73$.

13. f has crosses the x-axis at $x \approx -2.62$ and at $x \approx 2.25$, so these points are roots of f.

15. f has crosses the x-axis at $x \approx 0.74$ and at $x \approx 3.02$, so these points are roots of f.

17. f has crosses the x-axis twice, at $x \approx -1.91$ and $x \approx -0.671$, so these points are the only roots of f.

19. (a) A graph of $f(x) - r(x)$ reveals that $|f(x) - r(x)| \leq 1/2$ if $-1 \leq x \leq 1$.

Copyright © Houghton Mifflin Company. All rights reserved

(b) From a graph of $f(x) - r(x)$, one can determine that $|f(x) - r(x)| \leq 0.001$ if $-0.32 \leq x \leq 0.32$. Thus, $a = 0.32$.

21. A graph of $f(x) - g(x)$ reveals that $-0.033 \leq f(x) - g(x) \leq 0.018$ if $1 \leq x \leq 3$.

23. The viewing window $[1.9, 2.1] \times [3.6, 4.4]$ works.

25. The viewing window $[-0.05, 0.05] \times [0.9, 1.1]$ works.

27. The viewing window $[4.995, 5.005] \times [0.9, 1.1]$ works.

29. The viewing window $[-0.01, 0.01] \times [0.9, 1.1]$ works.

31. The viewing window $[0.9, 1.1] \times [0.9, 1.1]$ works.

33. The viewing window $[-0.1, 0.1] \times [0.9, 1.1]$ works.

35. $-1 \leq f(x) \leq 1$ for any x. Furthermore, $f(\pi/2) = 1$ and $f(-\pi/2) = -1$. Since $\pi/2 \approx 0.157$, the maximum value of f over the interval $[-10, 10]$ is 1; the minimum value is -1.

37. It is clear from a graph that f achieves its maximum and minimum values at the left and right endpoints of the interval $[-10, 10]$, respectively. Thus, the maximum value of f over the interval $[-10, 10]$ is $f(-10) = 10/21 \approx 0.476$; the minimum value is $f(10) = -10$.

39. The maximum value of f over the interval $[-10, 10]$ is $f(0) = 0$; the minimum value is $f(10) = -100$.

41. The maximum value of f over the interval $[-10, 10]$ is $f(10) = 100$; the minimum value is $f(0) = 0$.

43. The maximum value of f over the interval $[-10, 10]$ is $f(10) = 924$; the minimum value is $f(-10) \approx -99.999$.

45. The maximum value of f over the interval $[-10, 10]$ is $f(0) = 3/4$; the minimum value is $f(10) = 3/104$.

47. The maximum value of f over the interval $[-10, 10]$ is $f(10) = 11^3 = 1331$; the minimum value is $f(-10) = (-9)^3 = -729$.

49. The maximum value of f over the interval $[-10, 10]$ is $f(10) = 10{,}503$; the minimum value is $f(0) = 3$.

51. The maximum value of f over the interval $[-10, 10]$ is 72; its minimum value is -72. (The maximum and minimum values occur at $\pm\sqrt{72} \approx \pm 8.485$.)

53. For any x, $-1 \leq \sin x \leq 1$. If $-1 \leq x \leq 1$, $\cos 1 \leq \cos x \leq 1$. Thus, the maximum value of f over the interval $[-10, 10]$ is $f(0) = 1$; the minimum value is $f(\pi/2) = \cos 1 \approx 0.540$.

Appendix B

1. (a) This is the interval $-3 \leq x < 2$ (only the left endpoint is included in the interval).

   (b) This is the interval $-3 < x < 2$. (neither the right nor the left endpoint is included in the interval).

   (c) This is the interval $-3 \leq x \leq 2$. (both the right and the left endpoints are included in the interval).

   (d) This is the interval $-3 < x \leq 2$. (only the right endpoint is included in the interval).

3. The set defines an interval of length 10 that is centered at -1 and includes only the left endpoint. Thus, the interval is $[-6, 4)$.

5. The set defines an interval of length 7 that is centered at $11/2$ and includes both endpoints. Thus, the interval is $[2, 9]$.

7. (a) The interval $[-5, 3]$ has length 8 and is centered at -1. Since both endpoints are part of the interval, the interval is the solution set of the absolute value inequality $|x + 1| \leq 4$.

   (b) The midpoint of the interval $[a, b]$ is $(a+b)/2 = ((-5) + 3)/2 = -1$.

   (c) The radius of the interval $[a, b]$ is $(b-a)/2 = (3 - (-5))/2 = 4$.

9. x is a solution of the inequality if $2x + 3 \geq 5$ **or** $2x + 3 \leq -5$. Now, $2x + 3 \geq 5 \implies 2x \geq 2 \implies x \geq 1$. Also, $2x + 3 \leq -5 \implies 2x \leq -8 \implies x \leq -4$. Therefore, the solution set of the absolute value inequality, expressed in interval notation, is $(-\infty, -4] \cup [1, \infty)$.

11. The solution set consists of those points that are farther from 2 than from -3. In interval notation, the solution set is $(-\infty, -1/2)$.

13. The solution set consists of those points whose distance from -3 is greater than or equal to their distance from 2. In interval notation, the solution set is $[-1/2, \infty)$.

15. The absolute value inequality $|x - 11| \leq 0.02$ is equivalent to the double inequality $-0.02 \leq x - 11 \leq 0.02$. From this it follows that $10.98 \leq x \leq 11.02$, so $L = 10.98$ and $U = 11.02$.

17. The double inequality $-3 \leq x \leq 9$ describes an interval of length 12 centered at 3. Since the interval is closed, the interval is the solution set of the inequality $|x - 3| \leq 6$.

19. The double inequality $-7 \leq x \leq -4$ describes an interval of length 3 centered at $-11/2$. Since the interval is closed, the interval is the solution set of the inequality $|x + 11/2| \leq 3/2$.

21. The interval has length 10 and is centered at 5; it is an open interval. Therefore, the interval is the solution set of the (strict) absolute value inequality $|x - 5| < 5$.

23. The interval has length 6 and is centered at 1; it is an open interval. Therefore, the interval is the solution set of the (strict) absolute value inequality $|x - 1| < 3$.

25. The interval has length 10 and is centered at 2; it is a closed interval. Therefore, the interval is the solution set of the absolute value inequality $|x - 2| \leq 5$.

27. The inequality $L \leq x \leq U$ means that x lies in the interval $[L, U]$. This interval has *midpoint* $(L + U)/2$ and *radius* $(U - L)/2$ (draw a picture to convince yourself). Thus the original inequality is equivalent to the absolute value inequality
$$|x - (L + U)/2| \leq (U - L)/2.$$

29. The distance between the points is $\sqrt{(x_2 - x_1)^2 + (y_2 - y_1)^2} = \sqrt{(3-1)^2 + (4-2)^2} = \sqrt{8} = 2\sqrt{2}$.

31. The distance between the points is $\sqrt{(5-(-2))^2 + (2-(-5))^2} = \sqrt{98} = 7\sqrt{2}$.

33. Since M is the midpoint of the segment joining P and Q, $M = ((x_1 + x_2)/2, (y_1 + y_2)/2)$. Therefore,

$$d(M, P) = \sqrt{(x_1 - (x_1 + x_2)/2)^2 + (y_1 - (y_1 + y_2)/2)^2}$$
$$= \sqrt{\left(\frac{x_1 - x_2}{2}\right)^2 + \left(\frac{y_1 - y_2}{2}\right)^2}$$

and

$$d(M, Q) = \sqrt{(x_2 - (x_1 + x_2)/2)^2 + (y_2 - (y_1 + y_2)/2)^2}$$
$$= \sqrt{\left(\frac{x_2 - x_1}{2}\right)^2 + \left(\frac{y_2 - y_1}{2}\right)^2}$$

The desired conclusion follows from the fact that $(x_2 - x_1)^2 = (x_1 - x_2)^2$ and $(y_2 - y_1)^2 = (y_1 - y_2)^2$.

35. The circle with radius r and center (a, b) is described by the equation $(x - a)^2 + (y - b)^2 = r^2$. Thus, $(x - 1)^2 + (y + 2)^2 = 9$ is an equation of the circle with radius 3 and center $C = (1, -2)$.

37. Completing the square:

$$3x^2 + 3y^2 + 4y = 7 \iff x^2 + y^2 + 4y/3 = 7/3$$
$$\iff x^2 + (y + 2/3)^2 = 25/9$$
$$\iff \sqrt{x^2 + (y + 2/3)^2} = 5/3.$$

Thus the equation represents the circle with radius $5/3$ and center $(0, -2/3)$.

39. Most of the work is done in Example 7 of this section. (See the picture there.) All that remains is to see that the distance from m_1 to m_2 is $a/2$. A look at the coordinates of m_1 and m_2 shows that this is so.

41. The set is a closed interval with radius 5 centered at 2. Therefore, in interval notation, the set is $[-3, 7]$.

43. The set is an open interval with radius b centered at a. Therefore, in interval notation, the set is $(a - b, a + b)$.

45. There are various ways to do this. Here's a way that involves solving a quadratic equation. (Another—similar—approach uses the Pythagorean rule.)

The points we want are of the form $(x, -3)$, for some unknown values of x. Which values of x work?

The fact about distance says that

$$13 = \sqrt{(x - 1)^2 + (-3 - 2)^2} = \sqrt{(x - 1)^2 + 25}.$$

Squaring both sides gives the quadratic equation $x^2 - 2x - 143 = 0$. To solve this, either factor or use the quadratic equation; the result is that $x = 13$ or $x = -11$.

Conclusion: The two points we're looking for are $(13, -3)$ and $(-11, -3)$.

47. The interval between -5 and 3 has length 8 and center -1. The desired set is the points that are *not* part of this interval. Thus, the given set is the solution of the absolute value inequality $|x + 1| > 4$.

49. The given set can also be described by the double inequality $-7 < x < 5$. This is an interval of length 12 with center at -1. Therefore, the set is the solution of the absolute value inequality $|x + 1| < 6$.

51. Let T denote the temperature in the room. Then, $|T - 100|$ is how close T is to $100°$ F and $|T - 32|$ is how close T is to freezing. Therefore, the given sentence is equivalent to the inequality $|T - 100| < |T - 32|$.

APPENDIX B: REAL NUMBERS AND THE COORDINATE PLANE

53. For any r, $|r| \geq 0 \implies |r| \geq -2$ is a true statement.

55. If $3 < r < 7$, then If $r > 0$, $|r| = r$. Therefore, $r > 3 \implies |r| > 3$ is a true statement.

57. No. The hypothesis $-3 \leq x \leq 11$ allows the possibility that $x = 10$, a possibility that violates the given inequality.

59. No. The hypothesis $-3 \leq x \leq 11$ allows the possibility that $x = 0$, a possibility that violates the given inequality.

61. Yes. The hypothesis $-3 \leq x \leq 11$ implies that $|x| \leq 11$. Thus, the given inequality must be true.

63. No. The hypothesis $-3 \leq x \leq 11$ allows the possibility that $x = 0$, a possibility that violates the given inequality.

65. No, the hypothesis $-3 \leq x \leq 11$ means that $x = 11$ could be true. However, this value of x violates the given inequality. [NOTE: The double inequality $-3 \leq x \leq 11$ is equivalent to the absolute value inequality $|x - 4| \leq 7$.]

67. The inequality $|s| \leq 1$ is not true for all values of s that satisfy the inequality $-2 \leq s \leq 1$. For example, $s = -3/2$ satisfies the condition that $-2 \leq s \leq 1$, but $|s| = 3/2 > 1$.

69. For any number s, $|s| \geq 0$.

71. We'll work with inequalities:

$$|x - 3| \leq 0.005 \iff 2.995 \leq x \leq 3.005$$
$$|y - 2| \leq 0.003 \iff 1.997 \leq x \leq 2.003$$

Adding the last two inequalities gives

$$4.992 \leq x + y \leq 5.008,$$

or, equivalently,

$$|(x + y) - 5| \leq 0.008.$$

Appendix C

1. Let (x_1, y_1) and (x_2, y_2) be points on a vertical line. Since the line is vertical, $x_1 = x_2$. Therefore, the denominator of the expression for the slope of a line $(x_2 - x_1)$ will be zero. This means that the slope is undefined.

3. (a) Since the y-coordinate of both points is the same, the line is horizontal. Thus, the line is described by the equation $y = 3$.
   (b) The slope of the line is $(3 - 3)/(1 - (-2)) = 0$.
   (c) The line intersects the y-axis at the point $(0, 3)$. This point is the y-intercept of the line ℓ.
   (d) The line ℓ does not intersect the x-axis, so it does not have an x-intercept.

5. (a) If $A = 0$, the equation becomes $By = C$ or, equivalently, $y = C/B$ — the equation of a horizontal line.
   (b) If $B = 0$, the equation becomes $Ax = C$ or, equivalently, $x = C/A$ — the equation of a vertical line.
   (c) If $A \neq 0$ and $B \neq 0$, the equation can be written in the form $y = (-A/B)x + C/B$ — the equation of a line with slope $-A/B$.

7. Since $A = (1, 3)$ and $C = (6, 2)$, the slope of the secant line through these two points is $(2 - 3)/(6 - 1) = -1/5$. The equation of the line that passes through the point $(1, 3)$ with slope $-1/5$ is $y = (-1/5)(x - 1) + 3 = -x/5 + 16/5$.

9. Since $C = (6, 2)$ and $D = (9, 3)$, the slope of the secant line through these two points is $(3 - 2)/(9 - 6) = 1/3$. The equation of the line that passes through the point $(6, 2)$ with slope $1/3$ is $y = (1/3)(x - 6) + 2 = x/3$.

11. The slope of the secant line is $(3 - \sqrt{8.9})/(9 - 8.9) = 10(3 - \sqrt{8.9})$. Thus, the secant line is described by the equation $y = 10(3 - \sqrt{8.9})(x - 9) + 3 = 10(3 - \sqrt{8.9})x + (90\sqrt{8.9} - 267)$.

13. The slope of the secant line is $(\cos(1.5) - \cos(1.3))/(1.5 - 1.3) = 5(\cos(1.5) - \cos(1.3))$. Thus, the secant line is described by the equation $y = 5(\cos(1.5) - \cos(1.3))(x - 1.5) + \cos(1.5)$.

15. (a)

x	-10	-3	0	1	5	7
$f(x)$	0.544	-0.141	0	0.841	-0.959	0.657

(b)

Δx	7	3	1	4	2
Δy	-0.685	0.141	0.841	-1.80	1.62
$\Delta y/\Delta x$	$0.-098$	0.047	0.841	-0.450	0.81

(c) The values of $\Delta y/\Delta x$ are not (even approximately) constant.

(d)

x	1.35	1.37	1.40	1.41	1.43	1.47
$f(x)$	0.976	0.980	0.985	0.987	0.990	0.995

(e)

Δx	0.02	0.03	0.01	0.02	0.04
Δy	0.004	0.0054	002	0.003	0.005
$\Delta y/\Delta x$	0.209	0.185	0.165	0.150	0.120

APPENDIX C: LINES AND LINEAR FUNCTIONS

(f) The values of $\Delta y/\Delta x$ are (very!) approximately constant.

17. The cosine graph resembles the line that passes through the point $(1, \cos(1)) \approx (1, 0.5403)$ with slope $(\cos 1.01 - \cos 0.99)/(1.01 - 0.99) \approx -0.8415$. An equation of this line is $y = -0.8415(x - 1) + 0.5403$.

19. $y = 12(x - 2) + 8$ The graph $y = x^3$ resembles the line that passes through the point $(2, 8)$ with slope $(2.01^3 - 1.99^3)/(2.01 - 1.99) \approx 12$. An equation of this line is $y = 12(x - 2) + 8$.

21. The slope of L is $(0 - 2)/(-2 - 1) = 2/3$.

23. Since L passes through the point $(1, 2)$ with slope $2/3$, L can be described by the equation $y = \frac{2}{3}(x - 1) + 2$.

25. No. L passes through the point $(\pi, 2\pi)$ if and only if $(\pi, 2\pi)$ is a solution of the equation $y = 2(x - 1)/3 + 2$. Since it is not, L does not pass through this point.

27. (a) The given information implies that the line is described by the equation $y = 2(x - 2)/3 + 1$.

　　(b) The perpendicular line has slope $-3/2$. Therefore, it is described by the equation $y = -3(x - 2)/2 + 1$.

29. Lines parallel to the y-axis are vertical. The vertical line that passes through the point $(2, 4)$ is $x = 2$.

31. The line $y = 5x + 7$ has slope 5, so lines perpendicular to it have slope $-1/5$. The line with slope $-1/5$ that passes through the point $(-3, 1)$ is described by the equation $y = -(x + 3)/5 + 1 = -x/5 + 2/5$.

33. (a) The x-intercept of a line is the point where $y = 0$. Thus, the x-intercept is $(a - b/m, 0)$.

　　(b) The y-intercept of a line is the point where $x = 0$. Thus, the y-intercept is $(0, b - ma)$.

　　(c) In point-slope form, the equation of the line is $y = mx + (b - ma)$.

35. If $k \neq 0$, the equation $2x + ky = -4k$ is equivalent to the equation $y = (-2/k)x - 4$. Thus, the line will have slope m is $k = -2/m$.

37. If $k \neq 0$, the equation $2x + ky = -4k$ is equivalent to the equation $y = (-2/k)x - 4$. Thus, in this case, the line has slope $-2/k \neq 0$. If $k = 0$, the equation becomes $2x = 0$ so the solution of the equation is just the point $(0, 0)$. Thus, there is no value of k for which the equation describes a horizontal line.

Appendix D

1. $(x+3)(\pi x^{17} + 7) = x \cdot \pi x^{17} + 7x + 3\pi x^{17} + 21 = \pi x^{18} + 3\pi x^{17} + 7x + 21$

3. Yes, it is an expression that involves only the sum of constants and constant multiples of positive interger powers of x.

5. $T(w) = 3 + 4w = 3 + 4w^1$. Since the highest power of the variable is 1, this is the degree of the polynomial.

7. Since the highest power of the variable is 5, this is the degree of the polynomial.

9. Since the highest power of the variable is 123, this is the degree of the polynomial.

11. No. If $x + 2$ were a factor of p, then p could be written in the form $p(x) = (x+2)q(x)$, where q is a cubic polynomial. This implies that $x + 2$ is a factor of p if and only if $p(-2) = 0$. However, since $p(-2) = -24 \neq 0$, $x + 2$ is not a factor of p.

13. (a) The degree of p is the sum of the degrees of the terms. Thus, the degree of p is
    $1 + 1 + 2 + 1 + 1 + 3 \cdot 9 = 33$.

    (b) The graph of p crosses the x-axis wherever the graph of one of the terms in the definition of p crosses the x-axis. That is, p crosses the x-axis at $x = 0$, $x = 2$, $x = 6$, $x = -8$, and $x = 1$. Thus, the graph of p crosses the x-axis 5 times.

    (c) p has a root at x if $p(x) = 0$. Using the factored form of p, it is clear that p has roots at $x = -8$, $x = 0$, $x = 1$, $x = 2$, and $x = 6$.

15. $p(x) = (x + 16)(x - 2)(x - 1/3)$

17. This problem is takes some care. In particular, it's important to read the graphs carefully to find the locations of roots.

    (a) Graph C $\leftrightarrow$ (i); Graph B $\leftrightarrow$ (ii); Graph A $\leftrightarrow$ (iii); Graph D $\leftrightarrow$ (iv);

    (b) The quadratic polynomial shown in Graph C has *roots* at -4 and 1, so $(x + 4)$ and $(x - 1)$ are *factors*. Therefore, p must have the form $p(x) = k(x-1)(x+4)$ for some constant k. To find the value of k, notice from the graph that $p(0) = -4 = k(-1)(4)$. Thus, $k = 1$ and $p(x) = (x - 1)(x + 4)$.

    (c) Since the roots are at -1, 1, and 2, and the polynomial is cubic, it can be written in the form $q(x) = k(x+1)(x-1)(x-2)$. To find k, note from Graph D that (apparently) $q(0) = 2$; from this it follows that $k = 1$. Thus, the equation is $q(x) = (x+1)(x-1)(x-2) = x^3 - 2x^2 - x + 2$.

    (d) From Graph A we see that -1 is the root in question. Therefore
    $$p(x) = (x + r)(x^2 + s) = (x + 1)(x^2 + s),$$
    where s is still to be found. Since $p(2) = 15$,
    $$15 = (2 + 1)(2^2 + s) = 3(4 + s),$$
    so $s = 1$. Therefore, $p(x) = (x+1)(x^2+1)$.

    (e) The function can be written in the form $p(x) = ax^2 + bx + c$. Plug in the points $(0, 2)$, $(-3, 5)$ and $(2, 10)$ to find the coefficients a, b and c. After some algebraic manipulation the result is $p(x) = x^2 + 2x + 2$.

    Another approach is to see that the vertex is at $(-1, 1)$, so the equation is of the form $y = k(x + 1)^2 + 1$, for some k. Since $p(0) = 2$ (see the graph) it follows that $2 = k + 1$, or $k = 1$. Thus $p(x) = (x + 1)^2 + 1 = x^2 + 2x + 2$.

APPENDIX D: POLYNOMIALS AND RATIONAL FUNCTIONS

19. (a) Multiplying out $(x+a)^3$ and simplifying gives $x^3 + 3ax^2 + 3a^2x + a^3$, as desired.
    (b) If $p(x) = x^3 + 3x^2 + 3x + 1$, then $p(-1) = -1 + 3 - 3 + 1 = 0$.
    (c) Long dividing $p(x)$ by $(x+1)$ gives $p(x) = (x+1)(x^2 + 2x + 1)$.
    (d) Fully factoring $p(x)$ gives $p(x) = (x+1)(x+1)(x+1) = (x+1)^3$.

21. Since $f(x) = (x-1)(x+2)$, $f(x) > 0$ if $x-1$ and $x+2$ have the same sign (i.e., both are positive or both are negative). Thus, $f(x) > 0$ if $x < -2$ or $x > 1$.

23. Since $(x-1)^2 > 0$ for all $x \neq 1$, $g(x) > 0$ if $x + 3 > 0$ and $x \neq 1$.

25. Since $(x+3)^2 > 0$ for all $x \neq -3$, $m(x) > 0$ if $x^2 - x - 4 > 0$ and $x \neq -3$. Now, $x^2 - x - 4 = \left(x - (1-\sqrt{17})/2\right)\left(x - (1+\sqrt{17})/2\right)$ so $m(x) > 0$ if $x < (1-\sqrt{17})/2$ or $(1+\sqrt{17})/2 < x < 3$ or $x > 3$.

27. $p(x) = x^3 - x^2 - 2x = x(x^2 - x - 2) = x(x+1)(x-2)$. From the factored form it is apparent that p has roots only at $x = -1$, $x = 0$, and $x = 2$.

29. $r(x) = x^4 - x^2 = x^2(x^2 - 1) = x^2(x-1)(x+1)$. From the factored form it is apparent that r has roots only at $x = -1$, $x = 0$, and $x = 1$.

31. $t(x) = x^4 - 2x^2 + 1 = (x^2 - 1)^2$. From the factored form it is apparent that t has roots only at $x = -1$ and $x = 1$.

33. $x^2 + 2x + 1 = (x+1)^2$ so the polynomial has only one root, at $x = -1$.

35. $x^2 + x + 1 = (x^2 + x + 1/4) + 3/4 = (x + 1/2)^2 + 3/4$. Thus, this polynomial has no (real) roots.

37. $x^2 + 2x = ax^2 + bx + c$ so $a = 1$, $b = 2$, and $c = 0$. Thus, the quadratic formula implies that this polynomial has roots at $\dfrac{-2 \pm \sqrt{2^2}}{2}$. That is, its roots are $x = -2$ and $x = 0$.

39. $x^2 + 2x + 2 = ax^2 + bx + c$ so $a = 1$, $b = 2$, and $c = 2$. Since $b^2 - 4ac = 4 - 8 = -4 < 0$, the quadratic formula implies that this polynomial has no real roots.

41. $x^2 + bx + c = ax^2 + bx + c$ so $a = 1$. Plugging into the quadratic formula yields the following expression for the roots of the polynomial: $\left(-b \pm \sqrt{b^2 - 4c}\right)/2$.

43. $x + \dfrac{1}{x} = \dfrac{x^2}{x} + \dfrac{1}{x} = \dfrac{x^2 + 1}{x}$.

45. $3x\left(1 + \dfrac{5}{x} + \dfrac{2}{x^2}\right) = 3x + 15 + \dfrac{6}{x} = \dfrac{3x^2}{x} + \dfrac{15x}{x} + \dfrac{6}{x} = \dfrac{3x^2 + 15x + 6}{x}$

47. $\dfrac{x-3}{x+5} + \dfrac{x+6}{x-1} = \dfrac{(x-3)(x-1)}{(x+5)(x-1)} + \dfrac{(x+6)(x+5)}{(x-1)(x+5)} = \dfrac{(x^2 - 4x + 3) + (x^2 + 11x + 30)}{x^2 + 4x - 5} = \dfrac{2x^2 + 7x + 33}{x^2 + 4x - 5}$

49. $x + \dfrac{1}{x} = \dfrac{x^2}{x} + \dfrac{1}{x} = \dfrac{x^2 + 1}{x}$. This function has no real roots because $x^2 + 1 > 0$ for all x.

51. $3x\left(1 + \dfrac{5}{x} + \dfrac{2}{x^2}\right) = \dfrac{3x^2 + 15x + 6}{x}$. Now, this function has a root if $3x^2 + 15x + 6 = 0$. Thus, using the quadratic formula, we conclude that the rational function has roots at $x = \left(-5 \pm \sqrt{17}\right)/2$.

53. $\dfrac{x-3}{x+5} + \dfrac{x+6}{x-1} = \dfrac{2x^2 + 7x + 33}{x^2 + 4x - 5}$. This rational function will have a root wherever the quadratic polynomial $2x^2 + 7x + 33$ has a root. However, since $7^2 - 4 \cdot 2 \cdot 33 < 0$, this polynomial has no real roots. Therefore, the rational function has no real roots either.

Appendix E

1. For any $b > 0$, $b^0 = 1$. Therefore, $64^0 = 1$.

3. $64^{1/2} = \sqrt{64} = 8$

5. $64^{2/3} = \left((64)^{1/3}\right)^2 = 4^2 = 16$

7. $16 = 4^2 \implies x = 2$

9. $8 = 2^3 = (4^{1/2})^3 = 4^{3/2} \implies x = 3/2$

11. $4\sqrt{2} = 4^1 \cdot 4^{1/4} = 4^{1+1/4} = 4^{5/4} \implies x = 5/4$

13. $(\sqrt{a})^2 = (a^{1/2})^2 = a^1$

15. $1 = a^0$

17. $(a^3/a^5)^{10} = (a^{-2})^{10} = a^{-20}$

19. $a^{-2}a = a^{-2} \cdot a^1 = a^{-2+1} = a^{-1}$

21. No—0 is another number that is its own square. Thus $b^0 = 0$ is also consistent with the given statement.

    It *does* follow from the calculation above that either $b^0 = 0$ or $b^0 = 1$. However, to show conclusively that $b^0 = 1$ takes a bit more work.

23. Assume $x \geq y$. Then, $\dfrac{b^x}{b^y} = \dfrac{\overbrace{b \cdot b \cdot \ldots \cdot b}^{x\text{-times}}}{\underbrace{b \cdot b \cdot \ldots \cdot b}_{y\text{-times}}} = \overbrace{b \cdot b \cdot \ldots \cdot b}^{(x-y)\text{-times}} = b^{x-y}$. If $x < y$, the argument is similar.

25. Why is $a^x b^x = (ab)^x$ for positive numbers a and b and a positive integer x? Here's why:

    $$a^x b^x = \overbrace{a \cdot a \cdot \ldots \cdot a}^{x\text{-times}} \cdot \overbrace{b \cdot b \cdot \ldots \cdot b}^{x\text{-times}} = \overbrace{ab \cdot ab \cdot \ldots \cdot ab}^{x\text{-times}} = (ab)^x.$$

    The middle equality is the crucial one—it's legal because multiplication of real numbers is commutative: we can rearrange the a's and b's any way we like.

27. $g(1) = 5 = Ae^{0.2 \cdot 1} = Ae^{0.2} \implies A = 5e^{-0.2}$

29. Recall that $b^c = d \iff \log_b d = c$. Therefore, $8^{2/3} = 4 \iff \log_8 4 = 2/3$.

31. Recall that $b^c = d \iff \log_b d = c$. Therefore, $10^{-4} = 0.0001 \iff \log_{10} 0.0001 = 10^{-4}$.

33. Recall that $\log_b c = d \iff b^d = c$. Therefore, $\log_2 7 = x \iff 2^x = 7$.

35. Recall that $\log_b c = d \iff b^d = c$. Therefore, $\log_5 1/25 = -2 \iff 5^{-2} = 1/25$.

37. Recall that $\log_b c = d \iff b^d = c$. Therefore, $\log_{64} 128 = 7/6 \iff 64^{7/6} = 128$.

39. For any $b > 0$, $b^{\log_b c} = c$. Therefore, $2^{\log_2 16} = 16$.

41. For any $b > 0$, $b^{\log_b c} = c$. Therefore, $10^{\log_{10} 7} = 7$.

43. For any $b > 0$, $\log_b(b^c) = c$. Therefore, $\log_5 5^3 = 3$.

45. $e^{2\ln x} = \left(e^{\ln x}\right)^2 = x^2$

Appendix E: Algebra of Exponentials and Logarithms

47. Recall that for any $b > 0$, $\log_b(c/d) = \log_b c - \log_b d$, $\log_b 1 = 0$, and $\log_b b = 1$. Therefore, $\ln(1/e) = \ln 1 - \ln e = 0 - 1 = -1$.

49. Recall that $\log_b c^d = d \log_b c$. Therefore, $\log_4 A^2 = 2 \log_4 A = 2 \cdot 2.1 = 4.2$.

51. Recall that $\log_b(cd) = \log_b c + \log_b d$. Therefore, $\log_4 16A = \log_4 16 + \log_4 A = 2 + 2.1 = 4.1$.

53. For any $b > 0$, $\log_b c = d \implies b^d = (b^{-1})^{-d} = c \implies (1/b)^{-d} = c \implies \log_{1/b} c = -d$. Therefore, $\log_{1/4} A^{-4} = -4 \log_{1/4} A = 4 \log_4 A = 8.4$.

55. (a) Rule (B.2) $\implies b^x/b^y = b^{x-y}$
    (b) Rule (B.3) $\implies (b^x)^r = b^{xr}$

57. $f(x) = 20e^{0.1x} = 40 \implies e^{0.1x} = 2 \implies 0.1x = \ln 2 \implies x = 10 \ln 2$

59. $f(x) = 20e^{0.1x} = 10 \implies e^{0.1x} = 1/2 \implies 0.1x = \ln(1/2) \implies x = 10 \ln(1/2) = -10 \ln 2$

61. $h(1) = 40 = 20e^k \implies 2 = e^k \implies k = \ln 2$

63. $j(0) = 5 = Ae^0 = A$. Since $A = 5$, $j(3) = 10 = 5e^{3k} \implies 2 = e^{3k} \implies 3k = \ln 2 \implies k = (\ln 2)/3$.

65. $\log_2(x+5) = 3 \implies 2^3 = x+5 \implies x = 8 - 5 = 3$

67. $5^{x-1} = e^{2x} \implies e^{(x-1)\ln 5} = e^{2x} \implies (x-1) \ln 5 = 2x \implies (2 - \ln 5)x = -\ln 5 \implies x = -\ln 5/(2 - \ln 5) \approx -4.12082$

69. Let $y = \log_b x$. Then, $b^y = x \implies \log_a b^y = \log_a x \implies y \log_a b = \log_a x \implies y = (\log_a x)/(\log_a b)$.

71. $y = \ln\left(\exp(x^2)\right) = x^2$ for all x.

73. $\exp(\ln x) = x$ for all $x > 0$.

75. $y = \exp(\frac{1}{2} \ln x) = \sqrt{x}$ for all $x > 0$.

Appendix F

1. $\cos(1.2) \approx 0.35$ because 0.35 is the x-coordinate of the point on the unit circle that corresponds to an angle of 1.2 radians.

3. $\tan(2.8) = \sin(2.8)/\cos(2.8) \approx -0.35$

5. $\sin(4.5) \approx -0.98$

7. $x \approx 0.52$ and $x \approx 2.6$

9. $P(t)$ and $P(t+\pi)$ are diametrically opposite points. Therefore, if $P(t) = (x, y)$, then $P(t+\pi) = (-x, -y)$.

11. If $P(t) = (x, y)$, then $\sin t = y$ and $\sin(t+\pi) = -y = -\sin t$.

13. The graph through the origin is that of the sine function.

15. $\tan x = (\sin x)/(\cos x)$, so $\tan x = 0$ if and only if $\sin x = 0$. This happens at all integer multiples of π.

17. For any x, $\sin(-x) = -\sin x$. Thus, $\sin(-a) = -\sin a = -b$

19. For any x, $\cos(-x) = \cos x$. Thus, $\cos(-c) = \cos c = 0.3$.

21. $\sin^2 a = (\sin a)^2 = b^2$

23. $\sin 0 = 0$.

25. $\tan 0 = \dfrac{\sin 0}{\cos 0} = \dfrac{0}{1} = 0$

27. $\cos(3\pi/2) = \cos(2\pi - \pi/2) = \cos(2\pi)\cos(-\pi/2) - \sin(2\pi)\sin(-\pi/2) = 1 \cdot 0 - 0 \cdot (-1) = 0$

29. $\cos(5\pi/6) = \cos(\pi - \pi/6) = \cos(\pi)\cos(-\pi/6) - \sin(\pi)\sin(-\pi/6)$
    $= \cos(\pi)\cos(\pi/6) + \sin(\pi)\sin(\pi/6) = -\dfrac{\sqrt{3}}{2} + 0 \cdot \dfrac{1}{2} = -\dfrac{\sqrt{3}}{2}$

31. $\tan(3\pi/4) = \dfrac{\sin(3\pi/4)}{\cos(3pi/4)} = \dfrac{\sqrt{2}/2}{-\sqrt{2}/2} = -1$

33. $\sec(5\pi/3) = \dfrac{1}{\cos(5\pi/3)} = \dfrac{1}{\cos(2\pi - \pi/3)} = \dfrac{1}{\cos(\pi/3)} = \dfrac{1}{1/2} = 2$

35. $\sin(257\pi/3) = \sin(86\pi - \pi/3) = \sin(86\pi)\cos(-\pi/3) + \cos(86\pi)\sin(-\pi/3)$
    $= 0 \cdot \cos(\pi/3) - 1 \cdot \sin(\pi/3) = -\dfrac{\sqrt{3}}{2}$

37. $\cos\left(\dfrac{5\pi}{12}\right) = \cos\left(\dfrac{\pi}{4} + \dfrac{\pi}{6}\right) = \cos(\pi/4)\cos(\pi/6) - \sin(\pi/4)\sin(\pi/6) = \dfrac{\sqrt{2}}{2} \cdot \dfrac{\sqrt{3}}{2} - \dfrac{\sqrt{2}}{2} \cdot \dfrac{1}{2} = \dfrac{\sqrt{6}-\sqrt{2}}{4}$

39. $\sin\left(\dfrac{\pi}{12}\right) = \sin\left(\dfrac{\pi}{4} - \dfrac{\pi}{6}\right) = \sin(\pi/4)\cos(-\pi/6) + \cos(\pi/4)\sin(-\pi/6) = \dfrac{\sqrt{2}}{2} \cdot \dfrac{\sqrt{3}}{2} - \dfrac{\sqrt{2}}{2} \cdot \dfrac{1}{2} = \dfrac{\sqrt{6}-\sqrt{2}}{4}$

41. Yes. For any x, $\sin^2 x + \cos^2 x = 1 \implies \cos^2 x = 1 - \sin^2 x$.

43. No. $\sin(x - \pi/2) = \sin x \cos(-\pi/2) + \cos x \sin(-\pi/2) = \sin x \cdot 0 - \cos x = -\cos x$

45. Yes. If $\cos x \neq 0$, then $\sin^2 x + \cos^2 x = 1 \implies \dfrac{\sin^2 x}{\cos^2 x} + 1 = \dfrac{1}{\cos^2 x}$. Since $\tan x = (\sin x)/(\cos x)$ and $\sec x = 1/(\cos x)$, the last equation is equivalent to $\tan^2 x + 1 = \sec^2 x$.

47. $\sin(s - t) = \sin s \cos(-t) + \cos s \sin(-t) = \sin s \cos t - \cos s \sin t$

49. $\sin(2t) = \sin(t + t) = \sin t \cos t + \cos t \sin t = 2 \sin t \cos t$

Appendix G

1. Notice that there are two runs of fence of length y and one of length x. Therefore, the cost information implies that $3 \cdot x + 2 \cdot 2 \cdot y = 3x + 4y = 900$, or, equivalently, $y = 225 - 0.75x$.

    (a) By the formula above, area $= x \cdot y = x \cdot (225 - 0.75x)$.

    (b) We must have both x and y non-negative, so $x \geq 0$ and $225 - 0.75x \geq 0$. Notice that $0.75x \leq 225 \implies x \leq 300$. Thus the domain of A is the closed interval $[0, 300]$.

    A is a *quadratic* function, with roots at $x = 0$ and $x = 300$. Hence the *maximum* value of A occurs at the vertex, which lies midway between the roots—i.e., at $x = 150$. (The corresponding value of y is $225 - 0.75 \cdot 150 = 112.5$ feet; thus the "best" plot has dimensions $150' \times 112.5'$.)

    The corresponding area, therefore, is $A(150) = 150 \cdot (225 - 0.75(150)) = 16875$ square feet.

    (c) With twice as much money we'd have $3x + 4y = 1800 \implies y = 450 - 0.75x$, so $A(x) = x(450 - 0.75x)$. As before, A is quadratic, this time with roots at $x = 0$ and $x = 600$, so the maximum occurs at $x = 300$. The maximum *area*, therefore, is

    $$A(300) = 300 \cdot (450 - 0.75 \cdot 300) = 67500 \text{ square feet.}$$

    Thus *doubling Zeta's money quadruples the area of Zeta's fence.*

    A couple of notes:

    ▸ In both cases Zeta spends half the money on "horizontal" fence and half the money on "vertical" fence.

    ▸ The problem can also be done by working with a **parameter**, say M, for the amount of money available.

3. (a) If x is the length of each cut, the completed box has depth x, "width" $100 - 3x$ and "height" $50 - 2x$. The volume, therefore, is the function

    $$V(x) = x \cdot (100 - 3x) \cdot (50 - 2x).$$

    (b) The volume function is best maximized by plotting V. A plot shows that the volume appears to take its largest value—around 21,058 cubic cm—if $x \approx 9.4$ cm.

5. Proportionality means that that $A = kB$ for some constant k. Now $A = 1$ and $B = 2$ means that $1 = 2k$, so $k = 1/2$, or $A = B/2$. Similarly, $A = m/C$, $A = 1$, and $C = 3$ imply that $1 = m/3$, so $m = 3$ and $A = 3/C$.

    (a) From $A = (1/2)B$ and $A = 3/C$ we see that if $A = 10$, then $B = 20$ and $C = 3/10$.

    (b) Note that $(1/2)B = A = 3/C \implies B = 6/C$. This means that B and C are inversely proportional.

    (c) If $B = 10$, then $A = (1/2)(10) = 5$ and $C = 6/B = 6/10 = 3/5$.

7. Consider a circle of radius r. Then, $d = 2r$ is the diameter of the circle and its area is $\pi r^2 = \pi (d/2)^2 = (\pi/4)d^2$. Thus, the area of a circle is proportional to the square of its diameter. (The constant of proportionality is $\pi/4$.)

9. We are told that if w is the weight of the object and d its distance from the center of the earth, then $w = \dfrac{k}{d^2}$. So if $w = 200$ and $d = 10000$ we get

    $$200 = \frac{k}{10000^2} \implies k = 2 \times 10^{10}.$$

(a) From above, $w = \dfrac{2 \times 10^{10}}{d^2}$.

(b) To get $w = 100$ we need $100 = 2 \times \dfrac{10^{10}}{d^2}$. Solving for d gives $d^2 = 2 \times 10^8 \implies d \approx 14142.1$ miles.

Notice that (b) says that to halve w we must multiply d by $\sqrt{2}$. This is consistent with the proportionality statement in the hypothesis.

11. $T(r) = r/4 + \sqrt{3^2 + (6-r)^2}/3$ when $0 \le r \le 6$.

13. Let r be the radius of the circular pond, S be the surface area of the pond, and V be the volume of water in the pond. Then, $S = \pi r^2$ and $V = \pi r^2/2$. Thus, $S = 2V$.

15. Let the numbers be x and y. Since $x + y = 45$, $y = 45 - x$. Therefore, $x^2 + y^2 = x^2 + (45 - x)^2$.

17. Assume that the side of the triangle parallel to the x-axis has length 4 m and that the side of the triangle parallel to the y-axis has length 12 m. Then, the third side of the triangle is the line $y = 12 - 3x$. This means that if $0 < x < 4$ is the length of the side of the fenced-in rectangular play zone on the x-axis, the area of the play zone is $A(x) = xy(x) = x(12 - 3x) = 3x(4 - x)$.

19. The noise intensity at a point z textm from its source is $I(z) = k/z^2$, where k is a constant of proportionality. We'll assume that $k = 1$ for the quieter machine, so that $k = 4$ for the noisier machine. Thus, x m from the noiser machine, the noise intensity is $I(x) = 4/x^2$. Now, a point x m from the noiser machine is $100 - x$ m from the other machine, so the total noise intensity function is
$I(x) = 4/x^2 + 1/(100 - x)^2$.